REBUILDING THE PLANET
Episode One: Gathering

REBUILDING THE PLANET,

Episode One: Gathering

ISAAC WOLFE

REBUILDING THE PLANET,
Episode One: Gathering

World Ahead Press is a division of WND Books. The views and opinions expressed in this book are those of the author and do not necessarily reflect the official policy or position or WND Books.

Paperback ISBN: 978-1-944212-60-5
eBook ISBN: 978-1-944212-61-2

Printed in the United States of America
16 17 18 19 20 21 LSI 9 8 7 6 5 4 3 2 1

Unless otherwise noted, scriptural quotations are taken from The Holy Bible, English Standard Version, ESV text edition: 2007, Crossway Bibles, Wheaton, IL (with the author's renderings for Hebrew names.)

INFORMATION PAGE

Rebuilding the Planet — Episode One: Gathering

Copyright © Isaac Wolfe Books

ISBN: 978-1-944212-60-5

Original art by Antonio Garcia of Latter Rain Designs and are the property of Isaac Wolfe Books. Photos by Isaac Wolfe Books.

Cover design, editorial and consultation services by Clint and Marcy Chapman of GreenParrotPress, Inc.

Hebrew and transliteration edits by Jeremiah Greenberg of Messianic Liturgical Resources.

Additional information online at -
www.RebuildingthePlanet.com

CONTENTS

PROLOGUE

WOE! WOE! WOE!

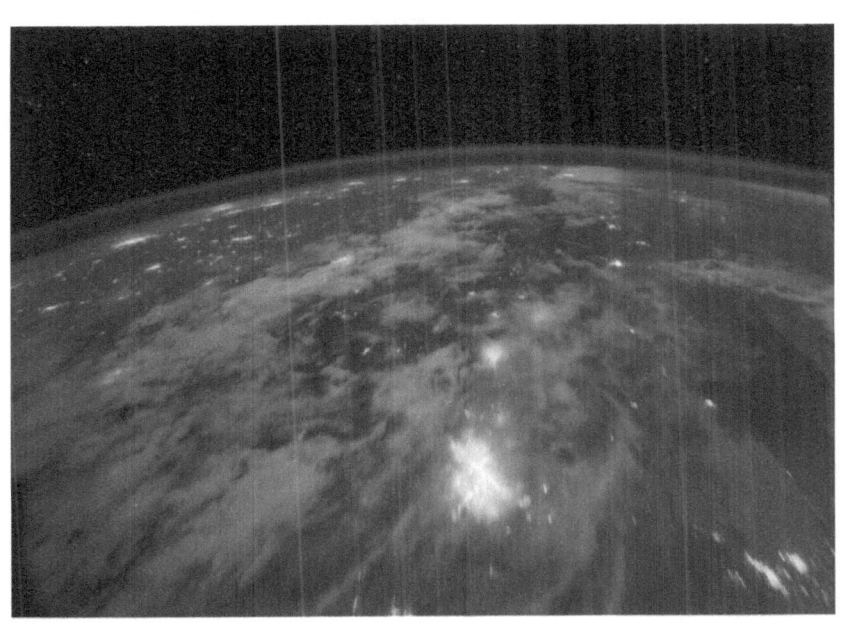

PLANET EARTH:

"The first messenger sounded, and there followed hail and fire, colored red, raining down upon the earth; a third of the trees were burned up, and the green grasses were burned up.

"Then the second messenger sounded, and a great rock mountain burning with fire, fell into the sea; and a third part of the sea was polluted; and a third part of the living creatures in the sea died; and a third part of the ships were destroyed.

"Then a third messenger sounded, and a great star from the sky, burning as a lamp, fell into the fresh waters into a third of the rivers and the fountains. That star is named Wormwood; and a third part of the fresh waters became wormwood; and many people died from the polluted waters.

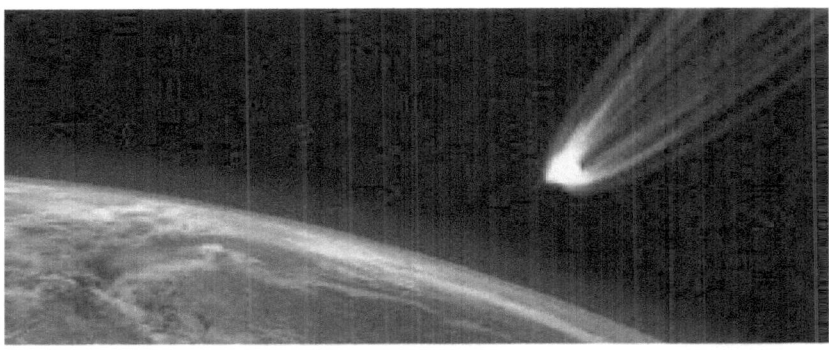

"The a fourth messenger sounded, and the sun was darkened by one third, as was the moon and the star; they were darkened by one third, and the daylight and the night lights.

"And I saw and heard another messenger flying though the sky, saying, with a loud voice, **'Woe, Woe, Woe,'** to the inhabitants of the earth, by reason of the sounds of the trumpets of three more messengers, which are yet to sound."

Revelation 8:7–13

CHAPTER 1

WISCONSIN

The planet Earth is being devastated.
Worldwide power upheavals,
Wars, and plagues are relentless.

"O' God . . . Please . . . somebody help . . . I can't give up. Must keep going," Isaac whispers, barely able to get the words out.

Marine Captain Isaac David Wolfe is only thirty-one years old, but he is sick and exhausted. He has survived scores of disasters, but just barely. He just does not know how much more he can take.

He clears his parched throat. "Guys! Anyone awake? Can any of you worthless jarheads help me up?"

Ellis raises his head and glances toward Isaac. "Yeah. Cool it, man. I'm awake. Just sorta dozed off for a bit. Hey, Ian, gimme a hand."

Ellis and Ian are also weak, but they struggle up to help their buddy. Aaron, a little stronger than the others, gets up by himself. Finally, all four slowly stumble about on the dirty floor, staggering toward the sheltered loading area of the old warehouse.

They move outside under the dark, moonless sky as remaining strength and hope drains from their weary bodies and minds.

It was over six years ago that their fifty-strong platoon of Marines had been redeployed back to the US to oppose the massive Gia Union invasion. Today, only four remain, Isaac, Aaron, Ian, and Ellis.

Suddenly, they are startled by a red radiance rapidly expanding in the eastern sky.

Ellis yells, "Look, the whole damn sky is lighting up! What the . . . ?"

The sky glows very brightly. Then just as quickly, it fades into a brilliant afterglow that rapidly dies away. The dark night returns with an unnerving, apprehensive silence lasting almost an hour. The weary men are watchful and more alert, trying to prepare for the unknown.

Without warning, the heavy black sky bursts back to life, filled with uncountable shimmering fiery trails. Molten rocks have been ejected by the crash of a large meteor in the North Atlantic Ocean.

Realizing incoming imminent danger, Aaron assumes command. "Quick, get cover! Back . . . in the basement!"

Helping Isaac, the group struggles into the lower level. At the last second, Aaron slams the heavy metal cellar door behind them.

Outside the sky is dazzling with incoming molten rocks, heard crashing as the flaming fragments slam to the ground. The building above the weary team is scorched and blackened, but none of the fiery projectiles break through into their reinforced basement.

Finally, Ellis ends a long silence. "Damn . . . seems like we have been down in this hell-hole for an hour."

"Actually, my guess would be two." Isaac's voice is faint and a little detached.

The air in the basement is steadily growing stuffy. Smoke begins to seep in and oxygen levels seem to be dropping further.

Ellis asks, "You guys hangin' in? I can hardly breathe. Maybe we should breathe more slowly, preserve our oxygen."

No one responds for several minutes. Ian says, "The pounding has stopped. You think maybe the damn storm's over?"

Aaron says, "Sounds clear to me and I need air too. I'm goin' up and hav' a look."

The others want to go as well, but Aaron is the strongest, with the fewest injuries and is the right choice. As he stands, he grabs Isaac's walking stick. Not wanting to touch the outer metal door with his bare hands, he uses the makeshift cane to push the partially blocked door open. As it swings wide, a blast of thick white choking smoke rushes in and quickly fills the room and basement. Three of them painfully fall to the floor, face-down, covering their mouths and noses with their shirts to filter out the smoke particles. Aaron covers his face with his dirty shirt hem and pushes forward into what has been their most recent shelter.

After a couple of minutes, Aaron steps back into the warehouse and yells down to his buddies, who are still on their faces in the basement. "Hell! It's a mess up here! But there's a strong wind kickin' up. I think it's from a blast-wave or somethin' from wherever that damn thing hit. Smoke's clearin', so you can come on up. Probably clearer up here than down there now."

Ian says, "Isaac's going to have a hell of a time getting out of here. Where's his stick?"

Aaron looks down at Isaac trying to get to his feet. "Gimme a sec, an' I'll help all your sorry asses out of this dungeon."

As he emerges, Ellis looks around at the smoldering area, still partially lit by small fires. "Son of a bitch! We're damn lucky this old building is concrete and reinforced steel. Look, our gear got some protection, just badly scorched . . . well actually now more *badly* scorched.

"Sure wonder what the hell's coming next. All these fricking catastrophes; all of 'em are merging together. I can't figure if it's over, or if there is more to come."

Isaac is responding to the fresher air and does not hesitate. "Ellis, you remember the angel's warnings! It was so dramatic I can't get it out of my mind. I remember every word, so you can bet your sweet ass there's a lot more to come."

"Yep, I agree." The other two Marines say, almost in unison.

Tired and in pain, but now with a charge of adrenalin, everyone is awake and alert with a temporary flash of energy.

Ian and Ellis take advantage of smoldering embers close by and build a campfire on the loading area. The four sit down on the hard dirty floor and lean back against the steel supports as they remember the series of disasters, earthquakes, firestorms and floods. They recall the signs from the sky months ago, signs, like right out of the Bible, Revelation-like prophecies and alarming words of warnings that were visually projected worldwide, like a three-dimensional hologram. It's hard to comprehend the magnitude of the events they have experienced.

Ellis uses a stick to poke at the comforting fire. "Isaac, can I take another look at your Bible? I wanna look up something I'm wondering about. I'm having trouble trying to figure out all this shit."

"Sure. Should be under what's left of my sleeping mat, with my journal and Siddur. While you're at it, bring me my journal."

"Roger that." Ellis successfully digs under the grimy, scorched pad and hands the journal to him.

Isaac says, "You know, that scruffy old Bible now seems quite relevant. But still, the Siddur seems kinda strange."

During a long reflection, Ian seems to be collecting his thoughts. "Isaac, I know you and Ellis are Jewish. So how did you end up having a Bible with a New Testament in it?"

"Long story. It's sorta tied to my Siddur, you know, my Jewish prayer book. Maybe you forgot, but my uncle was a rabbi. He gave me that Siddur when I graduated college and left home, you know, to go on active duty. He wanted to remind me how to pray. Well, that was his hope, but as you all know, I haven't used it a lot; at least not like it was supposed to be. My uncle expected me to pray two or three times a day. I didn't even come close. But I still like having it around. It's, kinda, well, kinda like a lucky charm. And maybe I might use it more someday."

Ian says, "If there *is* a someday."

Isaac adds, "Good point. But as to the Bible, its story is even more unusual. When my Aunt Ashley heard that my uncle had given me a Siddur, she slipped me this old Bible. It was very special to her since it's one she had used after she joined some sort of Messianic group many years ago. She said not to tell anyone . . . especially my uncle."

"And did you read it when she gave it to you?" Ian asks

Isaac pauses before replying, "Actually, I did. You know my passion for history. And I was anxious to read the history in it. Even though I had a history degree, I hadn't read an actual Bible. Of course, I always knew that all the miracles and stuff were just mythology, but it was still really interesting.

Ian frowns at Isaac. "Miracles? Mythology?"

"Yeah. We'd all learned in college that those things didn't really happen, thanks to good ol' Professor Frankenstein, as we called him. I even volunteered to do some archaeology with him one summer, digging at Tel Megiddo. That's the site of a lot of battles and wars. And the Bible makes a big to-do about it being the site of the last great battle in history.

"The ol' professor certainly added to our doubts about the Bible. He completely convinced us students that the stuff in the Bible was a collection of fables and myths, embellished by word of mouth transmission of itinerate peasants over the eons. His conclusions were confident and compelling."

Isaac stops, shakes his head and chuckles. "But I guess as we sit here today, we might want to reconsider the good ol' professor's 'certain and compelling' conclusions."

Aaron inquires, "Hell, I knew your dad was an astrophysicist and your mother was a history professor, but you also had a rabbi in the family . . . and they let you join the Marines?"

Ellis says, "You musta' been the black sheep or rebel of the family . . . I mean, along with your Messianic aunt!"

"No, no, it was all about money. You see, my parents didn't make that much. And with five kids, we all had to work out how to pay for college. So I joined the Marines. Hey, got my college paid for and some extra money to spend."

Ian asks, "Makes sense, but why'd you stay in? I thought you were always really *gung-ho*."

"Well, after a while, I just got into it. Sure didn't mean it to be a career, at least not at first. But ya' know, the adventure and the satisfaction gave me a feeling I was doing something kinda worthwhile. That kept me re-signing. Well hell, you gotta admit it has always been exciting and even fun, you know, until this Gia Union crap came along."

Isaac pauses, scoots a little closer to the warm fire, and gently massages his painful legs.

"My family wasn't very religious, but having a religious uncle, a rabbi, somehow seemed comforting.

"And since the Gia stuff started, I've been pulling Aunt Ashley's Bible out of the bag pretty regularly, readin' it over and over trying to dig deeper than just the history and stuff. Even though I still can't understand it all or, sorry to say, like knowing what's comin' next. You hafta admit, this old book has been pretty much on the 'mark,' to use a pun, about a lot of stuff we have been living through."

Ian says, "Yeah, I know. It was after you had re-read the last book of that Bible, that you convinced us to not accept the mark. Sure hope you were right. We committed a lot to that decision."

Aaron briskly turns to Ian. "Dammit, Ian. I can't believe I'm hearing you. Man, you're shootin' high and right! Damn! After all we've been through, disasters we've seen, and the hologram messages we've witnessed and you *'hope he's right.'* Good grief, man, he was right and still is!"

"Cool down man," Ian replies, "I know you're right and so's Isaac. Don't doubt that for a minute. It's just that if we don't survive all this, if this screwed-up world don't survive and the freaking human race is destroyed, it will not make a helluva lot of difference now, will it?"

With that, everyone gets quiet and drifts off into thought, generally about their dismal condition and how they seem to have exhausted their options.

Ellis flips the pages in Isaac's old Bible.

Isaac puts his Siddur and journal on his lap, squeezes his eyes closed tightly, trying to suppress the stabbing pains in his right leg, the burning from boils and draining, open wounds, and the deep throbbing pain of his broken left ankle.

Isaac wonders aloud, "How much more can we take? This sucks! At the rate we're going, I'm wondering if any of us will make it. And Ian may have a point. What's left to survive for? We're alone, can't find any of our families, and all that's left is either deserted or destroyed. It's so confusing. How could we be in this condition?"

Ian interjects, "Who knows why or what's going on? But I do know we can't stay here, we gotta find food and water. You're right. We're too weak. Somethin's gotta give."

Isaac looks up at his companions. "Guys, I think it's going to take a miracle to get us through this and I'm willing to keep trying until we find one, but we need some sleep before starting out again."

Ellis moves closer to Isaac. "I'm curious, you've been keeping a good log of all that's gone on haven't you? I'd hate to think that if none of us makes it, there's no record of what we've been through."

"Well, I started keeping the journal when I left home and did a pretty good job at first, though I have to admit, I've been pretty lax and off-an'-on since all the trouble started. But buddy, you're right . . . I wish I hadda' kept a better journal. It would just make me feel better if I knew that *our story* remained . . . even if we don't."

Ian says, "Still can't see what difference it makes, if there's no one else around to read it."

Ellis ignores Ian. "Well, I have an idea. Something maybe worthwhile if we don't survive, but others do."

"What?" Ian asks.

"Let's help Isaac update his journal, you know, compile our memories."

Everyone looks puzzled.

Aaron asks, "What are you talking about?"

Ellis concludes, "We will do it here, before we break camp. I can write pretty fast. So while we can, let's record our memories. If nothing else, it will keep us occupied mentally. And who knows, if we don't make it through all this, maybe a thousand years from now some young archeology student will find it and make it his doctoral thesis."

"And if we do make it?" Aaron asks.

"Then we can read it to our grandchildren."

"Grandkids?" Aaron exclaims.

All, except Isaac, smile at each other. He just quietly stares at his journal.

Ellis says, "C'mon, Isaac, whaddaya say?"

Isaac slowly says, "Well, I like you guys, but I'm not sure I'm ready to share my journal entries with you or anybody else. So come up with another idea."

"No, no. I didn't mean that," Ellis says. "We don't need to read the private parts of your diary. Hell, we already can guess most of what's in it, since we watched you do it! Just give us an idea of what you've covered of all this mess and we can then add to it."

Isaac pauses for a while. "First thought: pretty damn silly! Sittin' here hungry and thirsty, waitin' on another meteor strike or to be captured by Gia Troopers."

Isaac continues to stare at the fire. "Second thought: well, maybe for our grandkids. What the hell guys, why not? I'm in!"

Ian and Aaron finally go along with Isaac, but with less enthusiasm.

The men arrange themselves in a circle around the fire, instinctively keeping an eye out for Gia Troopers.

Isaac flips through his journal, summarizing pertinent information previously recorded.

The group then begins their written chronicle with Ellis recording the date, their names and rank, Major Aaron John Joseph, Captain Isaac David Wolfe, Lieutenant Ellis Daniel Coplan, and Gunnery Sergeant Ian Thaddeus Ryan, all of the United States Marine Corp.

Ellis turns to face Aaron. "Okay Major, I'm ready to start. You get first chance to kick this off."

The journaling begins ...

Aaron - "Well, we're now close to the seventh anniversary of the Gia Union. We have finally been able to conduct a recon of our former homes here in southern Wisconsin and, sadly, have found no one or any sign of what may have happened to them. We are jointly recording this from a warehouse on the bank of the Bark River and Crooked Lake. When we last saw this lake, it was a crystal-clear, spring-fed glacial lake . . . but not anymore. It's not safe to drink or even to wash in. And we're low on water and gotta find another source soon."

Ian - "We've been hiding from roaming bands that plunder whatever they can take. They pillage, rape, and murder. But even more, we have been hiding from Gia Union Troopers, who hunt for those wandering bands and for rebel groups like ours."

Isaac - "It was nearly seven years ago when we decided not to accept The Marck, demanded by the Gia Union. Well, you know, I guess The Marck logically made some sense.

It was intended to keep track of world resources, to allow proper allotment and to minimize waste and duplication. It was supposed to keep track of economic interactions, population movement, local and regional and central resources and needs. It was to be a key to the centralized new world government.

"But I had read about such a mark in the Bible's book of Revelation and it was just too disturbingly similar for me. Now not much scares me, but when I thought about the GPS capability that is embedded in The Marck, I just couldn't accept it. Gia Union bureaucrats knowing where I was at all times, and what I was doing? No way."

Aaron - "Of the seven of us that agreed with Isaac, now there's only four. Even if we don't survive, we still agree it was the right thing to do."

Ian - "Well, up to now the Gia Union Troopers were good at tracking everyone, including the pillaging bands that had The Marck. You see, The Marck was always linked to a GPS. They had no problem knowing the location of the bands. The problem was in getting to them. But from us, those Gia bastards ain't gettin' no GPS signals!"

As the Gia Union had grown to be a worldwide power, directed by central decision making and central control, it rapidly developed advanced scientific and technical programs and equipment. Many of those developments would have been labeled as wishful thinking, an impossible dream or science fiction, only a very few short years before. However, during the last two years all transportation networks and equipment, including the Gia Union's special spy hover units and their futuristic Sky Pods, have been disrupted by the firestorms, floods, hail, lightening and tsunamis. Earthquakes have destroyed bridges and dams and changed the course of many rivers and coastlines. Also, the electromagnetic pulse originating from the Wormwood comet severely disrupted their GPS systems.

The fire is growing weak, so the group takes a short break in their journaling effort to gather more wood from the piles of nearby debris. They slowly replenish the supply and take a few minutes to stoke the campfire.

Ellis has been quiet while taking dictation, but now comments, "So guys, thinking about it, maybe there has been a positive side to all these plagues."

Isaac looks across the fire. "I think you're stretching it Ellis. However, I think they now have given up on the bands. In fact, we haven't encountered one in months. I guess the Troopers' focus now is to find and eliminate groups like us, who have refused and escaped implantation of The Marck. That is, if there still are any other groups like us. We're probably public enemy number one. But, on the positive side, they have no way to track us."

Aaron, annoyed by drifting smoke, moves to the opposite side of the circle. He settles down and says, "Okay guys, time to get back to work while we can. Ian, you got some more for Ellis to write?"

"Sure do!"

Journaling ...

Ian - "Rebellious life on the run has been really hard. Really hard. In the Gia Union society, money, cash, change, checks, and so forth are of no purpose. All transactions are done electronically, via The Marck. Funds are immediately transferred from you to another if you are buying, or to you from another, if you are selling. It is implanted into the back of your right hand, or into your scalp at the hairline, and it is always accessible. No need for credit cards, debit cards, Social Security cards, insurance cards, licenses, or passports. The Marck is the all in one. And worse, The Marck is programmed to allow or forbid transactions. So there's always a Gia Command permission program governing every transaction, for every interaction."

Isaac - "That's why we've not been able to buy or sell or make any electronic or financial exchanges. For us there are no legal ways to get supplies! We can't even cross borders . . . officially that is. If any are left, we couldn't get across a toll road or bridge. Hotels, motels, commercial and public buildings have all been off limits. Even private homes have electronic tracers, just like restaurants, public toilets, and parks."

Aaron - "From what we heard some time ago, there were initially many all over the world who refused The Marck, just like we did. Even though it was difficult for Gia Union Troopers to find such groups, the hardships of just keeping alive outside of the Gia society convinced many to give up and to 'come home' to the Gia Union, with promise of clemency. Finally, they just took The Marck. But not us!"

Isaac - "It was seven and a half years ago, when the United Nations, the European Union, China, Russia, India, Japan, United Korea, the Arab League, the African Union, and the Latin American Confederation forces passed power and control to the Gia Union.

"And if that wasn't enough, wow! Were we surprised a month later when the NATO forces signed on with them. NATO! Those were guys we knew, fought and died with. NATO? No NATO anymore."

Ian - "We were fighting in Iran when NATO signed over their power. That move isolated the United States politically *and stripped it of its long-standing NATO allies, all except for Canada."*

While other countries succumbed to the Gia Union, the United States and Canada staunchly refused. Economic pressure was applied first, but unsuccessfully. By then, the United States had established a significant level of energy independence by fully developing its oil, gas, shale, fracking, and clean coal resource technologies, as well as its wind, wave, solar, and nuclear energy systems. In addition, all of this was about to be overshadowed with a budding, revolutionary secret cold-fusion energy program. The US had become the world's largest energy supplier and was unfettered by foreign energy and financial control. The US had huge trade surpluses and had even paid off its foreign debts. The country was rich and powerful, and thought it could, with Canada and a few others, stand up to the Gia Union.

Journaling …

Aaron - "Unfortunately, America had stopped investing in military expenditures and had been shifting resources to its growing social programs, infrastructure, and, of course, entertainment and leisure. All that plus lower taxes. Hard to believe, but apparently no one had even dreamed of a military response, not even our stupid military leaders."

Ian - "Yeah, I agree. To everyone's disbelief, there was a massive surprise invasion by the Gia Union. By air, ground, and sea, 4 million Gia Union Troopers crossed US and Canadian borders, swarmed the coasts and airdropped into our interior. Our small standing forces were quickly routed. Most citizens had given up their guns, so they could not resist."

Ellis - "At the onset of the Gia Union invasion, our unit was redeployed back stateside to help repel the second wave of invaders into the heartland. At first, in Chicago, we fought alongside Special Ops units, Rangers and Green Berets. Then we deployed to Cleveland. After that, we went south to defend Atlanta. The fighting was intense, but our lack of military resources and manpower resulted in series after series of setbacks and retreats."

Isaac - "Let's face it. Our leaders and military forces were overwhelmed. It was only a short time before our government heads had to make 'the' critical decision: use our nuclear and space arsenals or not. They refused to do so on our own territory.

"Then, there comes that damn stinking Settlement! Who could have imagined? The president and Congress, those puny-assed sons-a-bitches! They sued for a political settlement. That lousy agreement allowed Gia Union Troopers to be based within the United States and Canada!"

Aaron - "And even worse, it consented to an integration of both forces into the Gia Union military. All our units were to stand down. The Gia Union was allowed to set up their Gia Command Centers, placed strategically, throughout the two countries.

"It was surrender! A friggin' surrender! After centuries, the United States surrenders."

The agreement also gave the Gia Union access to America's vast and growing energy resources. Initially, the cash infusion from the accelerated sales of oil, gas, coal, minerals, timber, and other resources sparked a financial boom for many Americans. Millions quickly grew very wealthy and investments ballooned.

People rapidly began to acclimate to this new, tidy society. Many found in this new system those things they had long been wanting. Many political, cultural, and religious leaders even hailed the new order as God-sent. Prosperity. Organization. Efficiency. There was an open worldwide market for American and Canadian goods and resources. And, officially, there was no more war, at least for the USA.

America and Canada, though prosperous, had become effectively an occupied territory. They were now subjugated by Gia Union Troopers and financially dependent upon the Gia Union, even though not at first officially Gia Union member states. All Canadian and US citizens were, from the very start, strongly encouraged to have The Marck implanted. This allowed for full international travel, international trading, and full legal rights, since anyone with The Marck was immediately considered a Gia Union citizen. In addition, Americans were given the right to join the Gia Union Trooper forces. American colleges and universities became fertile recruiting grounds for the Gia Union Cadet program.

Over the next few months, the standardization of the Gia Union power and access to American resources continued to further promote global economic growth. The whole world was getting richer with cheap American and Canadian goods, including their oil, gas, coal, timber, metals, and advanced technologies. It was strongly suspected that revolutionary cold-fusion secrets were also compromised.

Journaling ...

Isaac - "All the so-called gains and improvements after the settlement were short-lived. In America, Canada, and some other small non-Gia-member nations, rapid stripping of resources, resource manipulation, fiscal restrictions, and threats of embargos by the Gia Union Command led to increasing submission by political leaders. I think the bottom line is that these so-called leaders, damn traitors, were also just paid off."

Ian - "Paid off by the corrupt Command until the decisions were final. The United States, Canada, Mexico, and Central America joined together as the new North American Federation, a full member state of the Gia Union. All decisions would then be made from the Supreme Gia Command in Jerusalem, the world's capitol.

Aaron - "Orders were signed to proceed with the immediate and full absorption of the US Department of Defense into the Gia Union Forces. No more US Army, Navy, or Air Force and no more United States Marines!"

Ellis - "That absorption included the mandatory requirement for the implantation of The Marck in all service men and women. We were just outside of Atlanta when the order came. Of our ten surviving unit members, only two, Gary and Stan, elected to take The Marck. The other eight of us went AWOL."

Ian - "For quite a while we were pinned down in Georgia, since travel was almost impossible. But our strategy was to be patient, avoid detection, get out of the Atlanta area, and start making our way north. It was slow going. Without the implanted Marck, we had no access to food, clothing, transportation, or any kind of buying or selling. We had to live off the land, staying off the main fairways and out of towns."

Aaron - "Life on the run may be difficult, but it is right in line with our training. Actually, we loved it at first. Another adventure. Along the way we met several other renegade groups, but agreed to keep separate and go our own *way. This time, we were the hunted, not the hunters. There were frequent close calls from Troopers. We had dwindling resources and no clear mission. We saw no endgame. It was really tough when we lost two more of our group, Carl and Nick, in a gunfight with Troopers just north of Columbus."*

Isaac - "All this was taking a toll on us. But we kept inching, actually winding our way further northward, not knowing what we were looking for and just surviving a day at a time.

"After many months of hiding, scavenging and barely staying alive, we were suddenly shocked by the sound from a first angel-messenger."

Ian - "Actually, a succession of angel-messengers announced a series of plagues. All of us, and probably everyone, saw them and heard the bone-chilling sounds and words. It was something like a hologram-scroll in the eastern sky. This happening initially raised hope, but then fear again."

Ellis - "Those messages were shocking. At first, it looked like they might mean things were going to change for the better, but just when we started to think that things could not be worse—they got worse! We lost Joseph in the second firestorm when he did not make it into the shelter in time. Ralph died the next week of smoke asphyxiation. Now, there are only the four of us left."

As the numbers of their unit had grown smaller, their friendships strengthened, evolving into a resilient bond. Objectives changed from offensive to defensive and then to just survival. Their individual identities began to merge into a singular entity determined to overcome the overwhelming adversity threatening their existence. However, even with grit and drive, years on the run had yielded deteriorating physical conditions, fatigue, and emotional stress.

With a collective realization that their chances for survival were diminishing, they developed a new, and they hoped not final, plan. They decided to push hard and, as rapidly as possible, work their way back to familiar home territory. There, on their own turf they would know where to try to look for food, water and shelter. More importantly, they hoped they might find family members and friends who had survived the Gia Union occupation and the prophesized disasters that were following.

With difficulty, over several months, they traveled northward through Kentucky and Illinois. As they crossed the boundary into Wisconsin an inexplicable feeling of warmth, safety and comfort briefly washed over them. But reality became a sobering antidote as the facts on the ground quickly restored their high level of anxiety and apprehension. Nothing back home was the same.

Journaling …

Ian - "All four of us had been wanting to get back to our homes to find out what had happened to our families and friends. Although it would be slower, we decided it would be safer to stick together as a unit and make our way to each man's hometown.

"Being closer to Racine, my hometown, we carefully headed there first. I couldn't resist and pushed out in front, more exposed than I should have been. We moved into the edge of the apparently abandoned city, then closer to Lake Michigan. I couldn't believe it when we got to my home. No one was there. Neither was the house. It was burned down. Only the foundation was intact, with pipes and exposed wire protruding through the concrete. There were no signs of life or personal effects. Nothing from my past. Remaining fragments of stone and glass were scattered toward the west.

It looked like there had been a strong blast from the east that took out the house. Maybe they escaped. O God, I hope so!"

Aaron - "Ian roamed through the property, kicking up debris and wreckage. Occasionally, he bent over to pick up a fragment. He stood still looking to the west, turned to us and yelled, 'Okay, men . . . on to Genese Lake!'"

Ellis - "I admit I was scared when we continued on to the lake where my family had a summer home. Again, we found no one. The house was intact, with exception of some damage to the roof and windows being blown out on the east side. The furniture was in place, but all of our personal belongings were missing, along with anything of value, including food, dishes, eating utensils, and electronics. It had obviously been stripped clean by somebody. No vehicles, but I could see the top of the mast of the sailboat about fifty feet off shore. There were no signs of family or messages. We all went room to room to see if there was anything written down, anywhere on the walls, any scratches or signs of struggle. None."

Isaac – "We were just about to leave when I noticed a dirty grey ceramic chard with scratches on the back. I picked it up and handed it to Ellis to read."

Ellis – "What a shock! The writing was fairly clear, '**Ellis, Next year in Jerusalem, Mom.**' But none of us could figure out. Was it something from an old Passover Seder from many years ago, or had Mom had more recently left a message, just in case. But that was it. We spent another two hours looking for more clues or more messages, but found nothing."

With a bit higher purpose, the group had then hurried on to Isaac's family home on Oconomowoc Lake. As they approached the home from the south side of the lake, it was clear that the house was gone, along with the garages and barn. All the buildings had been burned and the debris scattered. The docks and boat house were collapsed and submerged. A careful search through the debris revealed no writings, engravings, or messages of any kind. Not this time. The group spent the night there before moving on eastward toward Aaron's hometown in Waukesha. Their efforts there were also fruitless, yielding nothing. There were no clues and no sign of where the family and neighbors had gone, the same demoralizing results.

Nothing found so far could serve as a game-changer. Nothing to build a resistance or plan around. Only the one burned grey chard from Genese Lake, addressed to Ellis. Nothing else, and there was no certainty of its relevance.

Confounded by their inability to find anyone remaining in the areas where they had previously experienced so many good memories and security, they decided to move further inland in hopes of finding some clean water and food. By the time they reached the Bark River, they were sick, hungry and dehydrated. On the way, Isaac suffered a broken ankle when a store wall collapsed while he was scavenging for food. They were approaching desperation and were grateful when they found the abandoned warehouse on the banks of the river.

Journaling …

Ian - "Even in isolated and remote locations, the plagues left severe disturbances in the air, water, and land. The familiar landscape we knew so well was disrupted. There had been floods, then firestorms and then the earthquakes! Who ever heard of earthquakes in southern Wisconsin?

"Despite all, we, well at least some of us, have survived. We ducked into empty basements during the firestorms and ran into the open fields during the many earthquakes, just

as we did during the Great Earthquake. The ravages of these events have resulted in piles of bodies. Corpses, burnt, crushed, decaying — all around.

"We found bodies of Gia Union troopers along the roads and streets. It seems as if they just leave their dead to rot, like animals. But the towns and roads now are empty. It certainly looks like the end of all life. Maybe the Earth is being cleansed?"

Aaron - "Or maybe the end of the earth? The berries, grasses, grains and bark, the supplies of nature, of the woods and fields are gone. We are surviving by scavenging canned food we find in the ashes of burned-out stores and from abandoned Gia Union warehouses. But those resources are almost gone. There's no more fish in polluted Crooked Lake or even in the larger Upper and Lower Nemahbin lakes. The Bark River is running red with incinerated organic matter and blood-red pigment. Rarely, a dead fish, duck or beaver floats down the winding stream. No telling what poisons they contain.

"The Corps has taught us a lot, but nothing has prepared us for the end of the human race!

"Ellis, for the record, be sure to note that the four of us are just barely holding onto life. I'm in the best shape, but not by too much. We have no more food left and this may be it for us. This may be the end. Maybe we'll just fall into a stupor or delirium, or fade into an eternal unconsciousness."

Ellis's hand is beginning to cramp and his eyes are blurry. "Okay guys, I'm exhausted, but I think we've done a hell of a good job getting this much info recorded. I think we should quit for now. Besides, I really gotta take a leak!"

Aaron says, "Yeah, we probably have a lot of it covered. But man, do you really have to waste water?"

The group is more than a little surprised at Aaron's uncharacteristic attempt at humor.

Isaac's pain has relented some and he is in the mood to continue with his own reflections. He pulls his mat close, leans back against the steel support beam and flips through his earlier entries in his personal journal.

To no one in particular, he speaks quietly about thoughts flowing past. "You know, my boyhood days around here don't seem so far away or so long ago. It was a long time ago, but then, it seems to be only yesterday. I remember kayaking down this same Bark River with Grandpa John and Aunt Ashley. I remember paddlin' thru the narrow rapids, having to duck under the low hanging willow branches. I remember gliding effortlessly by quiet meadows and along the forest covered banks. I remember frightening some surprised Canadian geese and the wood ducks as we turned a sharp bend. We even stopped to watch the busy beavers repairing their dam on a tributary creek.

"And I remember Carroll College, over in Waukesha, and being so proud of graduating in my Marine uniform. From then on it was a Marine's life for me, fulltime."

Isaac looks around and sees that he is basically talking to himself. The others are trying to find comfortable spots and no one is listening. He really doesn't care. It feels good to reminisce and think about all that has happened since leaving college and starting his journal. Reminiscing about the good ol' days, more pleasant times.

His thoughts go back to his early Marine training. His platoon had trained intensely and purposefully for special assignments, rapid deployment, and hostage rescue. Many in the platoon were seasoned Marines while others, like Isaac, were new to combat. This was where Aaron, Ian, Ellis and Isaac first met.

It was remarkable that they had so much in common, including coming from southern Wisconsin. Every man in the platoon grew to know the other guys, all like brothers. They learned each person's strengths and weaknesses, their reaction times and their personal habits. They didn't talk family that much, as the platoon was their family.

It was not long until they were called into action with their first deployment to Venezuela. After decades of communist and fascist rule, the country witnessed a serious change in popular support for pro-democracy rebels. The rebel opposition won the election for the presidency, despite the full-out corruption of the incumbent leadership. That upset created a violent reaction by the government, which had also become the western hemisphere's chief supporter of terrorism. Following the election, the communist regime captured the newly elected opposition leader, along with his family, and put them on trial. They planned a public execution, in the plaza in front of the Caracas palace.

The US government made a daring decision to immediately intervene. Isaac's unit had been developing a plan for two weeks, and were ready to roll. It was a nighttime raid on the palace. They successfully rescued the opposition rebel leader and his entire family.

The Marine unit remained there for twelve weeks during the establishment of democracy and a transfer to a democratic government. Isaac was assigned to protect Isabella, the new leader's twenty-four-year-old daughter. They grew fond of each other. But geography, politics and war saw that a long-term relationship was not to be. They kept up correspondence for a few months, but never met again.

After Venezuela, they were deployed to Gaza to assist counterterrorism efforts required by problems imported from Iran. They didn't make many friends there, where all the girls had their faces covered. They were never sure if there was a pretty girl under there, or just another suicide-bomber. They mainly just blew stuff up and rounded up suspects to turn over to the Israelis.

The next few years were busy years for the Marines and Special Ops units. The world seemed to be full of small to medium size wars. They were breaking out all around the globe.

After Venezuela and Gaza, the United States was involved again in Bosnia. It was a cold winter with over two hundred inches of snow blowing and drifting. But the snow seemed to have kept the insurgents pinned down and by spring the Marines had them on the run.

Next the unit went back to Afghanistan. Such are repeat wars, repeating tasks everyone had long thought to be accomplished. Those wars were mostly clashes of small units, special operations type engagements, but it was the very kind of warfare they specialized in. Never was war declared and, seemingly, never was victory.

Other than Venezuela, victories were not personally satisfying. Their effects were of fleeting significance since political issues were not the unit's priority. Getting the job done and returning safely, with no one left behind, that was the prime directive.

There were additional deployments to the Horn of Africa, to Lebanon and then across the border into Aleppo, Syria.

Aleppo was somewhat emotional for Isaac, since his family had had friends who had been 1947 refugees from Aleppo. In fact, his uncle, the rabbi, had a copy of the *Aleppo Codex*, the so-called "Crown of Aleppo," on display in his library.

However, it was in the Iranian War where their unit took the most casualties. None of the two of their original squads were intact by then. Only twenty-four of their original platoon were available for their next deployment to Kurdistan when that order came to help old Kurdistan Peshmerga friends push the Islamic Caliphate back . . . back again.

It is while they were in Kurdistan that the Gia Union invaded the United States and Canada. Their squad was urgently deployed back stateside, to help defend against the invasion. Isaac and Aaron both thought it was too bad they couldn't have taken some Kurds back to help.

As Isaac's attention shifts back to their current situation, Ian and Aaron have now curled up on their sleeping mats and Ellis is leaning against the steel support, slumped over and asleep too.

Only Isaac is conscious. He has a fever and his vision is impaired. He can just see blurred distant shadows now and his mind is confused. But for a moment, he thinks of Isabella of Venezuela, and then flashes back in time to Susie, her bright face, in her low-cut gown at the Marine Ball. Was that four years ago? Or six or eight? He's not sure. Months fly by in his mind. The timeline of his life is speeding by in seconds. Now, for a moment, he can't quite remember how his Marine buddies happened to be with him here, near his home.

The celestial Seal-Plagues, followed by the Trumpet-Plagues are opening in his memory.

It is all so confusing. All is such a blur. His mind is foggy and he is not sure if he is awake or dreaming. He cannot move his left arm and has increasing pain in his neck and head. He rolls over and drifts into unconsciousness, the last to drift off.

Now finally, when it appears that all is over.
And it looks like all is lost,
And now, all hope is gone,
It is . . .

"Time's UP!!

CHAPTER 2

"TIME'S UP"

"In a moment, in the twinkling of an eye . . . "
The pivotal event in world history. It is . . .

"TIME'S UP" TIME!

As the seventh, and final, celestial trumpet sounds, the sky peels open to reveal a vast scene projected across the eastern sky. The brilliant hologram is simultaneously visible in all places worldwide. The four Marines, awakened by bone-chilling sound vibrations, see the three-dimensional display clearly.

The depicted panorama of the heavens over Jerusalem almost looks like a science-fiction movie. There are white horses with riders brandishing light-swords and laser-tipped spears. The visual action is accompanied by powerful, ear-pounding, deep vibrating sounds created by an angel with a golden trumpet.

Nowhere is the reality of what is happening unseen, obscured, or hidden from view, including Wisconsin, 6,100 miles away from Jerusalem. Isaac and his friends feel the excitement of the event to the core of their suddenly attentive minds.

The majestic riders are led by One who carries a banner, inscribed with words that can be clearly read, *"King of kings, Lord of lords."*

Miraculously, everyone sees the message in their own language.

The Messiah has returned! The One for whom the world has waited for thousands of years . . . The Messiah. It is he who Jews and Christians together have prayed for, hoped for and waited for for so long.

This time, rather than in a manger in a small first-century village, the whole world beholds the coming of the Messiah, as King.

Some call him Jesus, the Christ, others simply, the Messiah. The title in Hebrew is Meshiach. But he calls himself by his

Hebrew given name, Yeshua. It is a name derived from the Hebrew word for salvation. As in the Hebrew gospel written by Matthew, nearly 2,100 years ago, an angel had said to Mary (Miriam, in Hebrew), "You shall call his name Yeshua, for he shall save his people." Yeshua is a shortened version of Yehoshua, God's salvation, in Hebrew.

Finally, now, at the lowest point in human history, at a time when hope for the survival of mankind and for the planet appears to be lost, He returns. This time in power and glory.

"And all behold him."

DAY 1 JERUSALEM

Yeshua, the Messiah and an uncountable host of mighty angels, equipped with light-swords and laser-tipped spears, have just arrived from the eastern sky and are descending toward Jerusalem. Yeshua's army also includes Kadoshim, who are resurrected human beings. *Kadoshim* is a Hebrew term translated as "holy ones." It also can mean "saints," or "elect." They are the righteous human men and women from the present and past ages, now resurrected in glorious bodies.

All peoples of the Earth see them as they zoom over the eastern mountains of Edom, over the Dead Sea, the Jordan Valley and then across the Judean wilderness toward Jerusalem.

This is Day One of a new age. The new age is to last for one thousand years, and then beyond. This new age has been prophesied for thousands of years. It has been referred to as "The Kingdom Age," "The Kingdom of Messiah" or more simply as "The Millennium."

As the world witnesses this event, every surviving person on earth experiences the landing of Yeshua and many of his Kadoshim on the Mount of Olives!

Yeshua, wearing a scarlet robe, dismounts and stands beside his great white horse. He is wearing a golden crown, like that of the High Priest, and the crown has "Holiness to Yehovah" inscribed on the front. Two Kadoshim, with light-swords drawn and holding laser-tipped spears, stand beside him. Others spread out over the western slope of the mountain while thousands of others hover over the mountain and the Kidron Valley between the Mount of Olives and the Old City of Jerusalem.

Yeshua stands on the Mount of Olives and overlooks Jerusalem.

DAY 1 WISCONSIN

Isaac cannot help but reflect on these familiar shofar sounds, resembling those that he had heard back in synagogue on Rosh Hashanah. The sounds of the strong, deep vibrations and sounds of the Golden Trumpet, *"Tekea, Teruah, Sh'varim, Tekea Gidolah."* These are the same sounds, but now a thousand times more powerful.

At first, Ellis, Ian, and Aaron are simply stunned and speechless, not quite sure if this is real, a hallucination, or a dream. But finally, there is no mistaking what is happening a half world away. All four men are awake and alert. Although they are frail and febrile, their hearts race, their breathing is rapid and their hands and lips tremble. They have seen strange, miraculous holograms before, but now they are filled with wonder. This time, it seems to be a wonderful message. They do not fear. They just look up.

Finally, Isaac and Ellis manage a smile to each other. All are too weak to speak but are experiencing a powerful sensation of shalom (peace). The overwhelmed men breathe deeply and drift off into a much-needed sleep.

DAY 1 JERUSALEM . . . CONTINUED

Yeshua, the Kadoshim and angels cover the western slope of the Mount of Olives, overlooking Jerusalem, just across the deep Kidron Valley.

Minor resistance from a surprised Olivet Security Unit of Gia Union Troopers, is quickly brushed aside by an ultra-sonic blast of his voice. Then with another ultra-sonic blast across the Kidron Valley, the stones closing the Eastern Wall and Gate are vaporized.

This Gate has been of interest for millennia. It has long been called the "Eastern Gate" and sometimes the "Golden Gate." It has been a subject of psalms, religious songs, praise, and a popular metaphor for entering a new life in a new world. Those songs and metaphors, based on the prophecies about the Eastern Gate, have been well known since the writing of the Hebrew prophet Zechariah, 2,500 years ago. Zechariah 14:4 prophesies, "And his feet shall stand in that day upon the Mount of Olives, which is before Jerusalem on the east."

Because of this well-known prophecy, the Kidron Valley, between the Mount of Olives and the Eastern Wall of the Old City of Jerusalem, has been filled with graves, deliberately. The western slope of the Mount has served as a prestigious Jewish cemetery for wealthy and prominent Jews seeking to be as close to the Messiah as possible when he descends to the Mount of Olives, as Zechariah predicted.

On the other side, the western side of the Kidron Valley, as it ascends upward toward the walls of the Old City, is a large Muslim cemetery. This cemetery has also been deliberately located below the Eastern Gate and eastern City Walls.

There is a long-standing belief that walking over a grave will defile a person, rendering him unclean. Islamic reasoning is that by locating graves below the Eastern Gate, any Jewish messiah who might happen to land on the Mount of Olives, would defile himself if he crossed over these graves. Defiled, he would no longer be able to cross the Kidron Valley, enter the Eastern Gate or enter the Temple Mount. This considerable effort was to interfere with a Hebrew Messiah entering the Temple Mount and Temple.

To further hinder such a Jewish messiah, Ottoman Sultan Suleiman the Magnificent, in 1541, commissioned the current Old City walls and further ordered the Eastern Gate sealed. It has remained blocked since. This current Eastern Gate is slightly north of the original Eastern Gate, which was called the Shushan Gate in Second Temple times, but it has remained the prominent structure of the eastern wall for over four hundred years.

Whether the grave defilement theology is true or not, Yeshua at His coming is not hindered by any such human barriers. Yeshua stretches forth his right hand and a golden bridge materializes, spanning the Kidron Valley. The Mount of

Olives is now connected with the Eastern Gate of the Temple Mount by this Golden Bridge.

Yeshua and his followers mount and ride their white horses over the bridge and through the blast opening of the Eastern Gate, without ever touching the hundreds of graves on either side of the valley. The Golden Bridge solves that issue. And it is more than just coincidental that Yeshua's new Golden Bridge spans the Kidron Valley, starting and ending from the same sites as once did an ancient triple-arched bridge. That ancient bridge was destroyed over two thousand years ago.

During the Temple's existence, for a thousand years, a mysterious ritual had been conducted on the Mount of Olives. The ritual was tied to and essential to the ceremonies in the temple across the Kidron Valley. The strange and mysterious ritual, called the red heifer ceremony, was necessary for cleansing the people and priests involved in the temple worship services. The altar for sacrifice of the red heifer was located on the western slope of the Mount of Olives. To purify the temple and its altars, the ashes from the red heifer were transported over a bridge to the temple. And going the other way across that bridge were the sin offerings that had been sacrificed in the courtyard of the temple, They were transported from the temple altar to their place of burning and burial on the slope of the Mount of Olives. This site was called "outside the camp."

Having crossed the valley on the new Golden Bridge and now on the Temple Mount itself, Yeshua dismounts and stands on the Mount. He looks westward where, until recently, a temporary tabernacle had stood. The Mount is in ruins, and needs to be cleared of all the clutter left by Gog and his henchmen. It has also suffered the ritual defilements of the Man of Sin, Gog, Supreme Commander of the Gia Union and needs ritual cleansing as well.

This time, however, the ritual of cleansing will not be by water mixed with hyssop and ashes of a red heifer, but this final time by the Messiah himself. In fact, the mysterious ceremonies of the red heifer were a typology, pointing to this work of the Messiah to ultimately cleanse the temple, the people, and the land. Pointing ahead to this very time.

The physical cleansing of the mountaintop site is high priority and gets underway immediately. Angels, Kadoshim and Yeshua himself dive into this urgent task. After much debris has been cleared from the surface, Yeshua, collects living waters from the Gihon Spring and brings it to the Mount. He selects a spot one hundred fifty meters to the west and ten meters south of the newly opened Golden Gate. He pours these living waters onto that spot, marking the site for a restored altar. Kadoshim and angels then proceed, busily constructing the altar. By midday the altar is ready. This is the first step.

Plans for constructing a massive new temple also get underway on Day One. These include the foundations of the spectacular new millennial temple, the courtyards, the support structures, the gates, the Holy Place, the Holy of Holies and all their contents. This will, however, be a year-long project while whole Temple Mount undergoes massive changes.

When the initial physical and ceremonial cleansing works have been completed, a golden throne is set up between the site for the altar of sacrifice and the site of the Holy Place. Yeshua now sits down on his golden throne, facing the place soon to become the Holy of Holies that will house the new ark of the covenant.

And thus begins His one-thousand-year reign on Earth.

It is now the time of the evening prayers, three o'clock, and a loud voice thunders from the heavens, "This is my beloved Son, in whom I am well pleased."

As millions of angels and Kadoshim hover over the city, its valleys and mountains, the whole host shouts,

"Baruch Haba b'Shem Yehovah!"
"Blessed is He who comes in the name of Yehovah!"
"Yeshua, The Messiah reigns!!"
"Yeshua, The Messiah reigns supreme!!!"
"Hallelujah!"

CHAPTER 3

RESCUE

DAY 2: WISCONSIN

It is a cool autumn morning as the sun begins to rise dimly through the dirty yellow-brown southern Wisconsin sky. Isaac and his three buddies have been in a sound, much-needed, refreshing sleep. In spite of being dirty, smelling of sweat, and having open draining wounds, this is their best night's rest in years.

Isaac and Ellis are the first to awaken. They can speak, but with dry, crackling and muted voices.

"Mornin'," Isaac says. "Did ya see all that?"

Ellis nods. "Unreal!"

Isaac says, "Maybe it was a dream or hallucination or somethin'. Or maybe I've just finally lost it!" He goes on to describe in a few words what he remembers seeing.

Ellis smiles. "Nope . . . don't think you've lost it . . . yet. I think we all saw the same thing. Unless I'm missing something too, I'm betting on 'real,' though unbelievable."

As Aaron and Ian awaken, each shares the same questions and the same answers. Thankfully, they all are also experiencing that rare feeling of having had a long, restful night of deep sleep.

"And I'm really hungry," Ian says.

"Too bad guys, we had the last of our salvaged can goods yesterday," Aaron says.

All four stretch and try to move about to loosen up. Realizing they are not only out of food, but out of water, they settle back to the ground, taking weight off their painful legs, while deciding what to do next.

Isaac glances toward the glow of the rising sun, which is obscured by the dense polluted air. He seems to see something, a light that is distinguishable from the blush of the masked sun.

"What's that?" Isaac asks.

All turn and strain to see that distant, focal light. It is growing larger. With as much energy as he can muster, Isaac crawls toward to his buddies. Anticipating a new and unknown threat, he grabs his walking stick and pushes himself upright.

Ellis says, "It's comin' this way, movin' without sound . . . yeah, definitely comin' our way and fast! Not sure what it is, but boys . . . we may be about ready to buy the farm.' This thing's scaring the shit out of me."

The men did not know that shortly after his entry into Jerusalem, Yeshua sent thousands of Kadoshim to provide for, to feed, to water, and to comfort the surviving remnants of humanity. Yeshua directed them. "Comfort you, comfort you my people."

The approaching light is one Kadosh heading straight toward the four ragtag Marines, who seem to be frozen in place. He has flown to southern Wisconsin to find Isaac and his three frail and confused friends. As he arrives, landing directly in front them, they are paralyzed by fear and uncertainty, but now very alert. Their eyes are dry and crusty. Their lips are cracked by dehydration. Their skin is stiff, lacking normal turgor.

"Shalom," he says. He smiles broadly and enthusiastically gestures toward the young men, who are shocked! They stare, but cannot get a word out of their dry throats to reply.

Not surprised by their stunned reaction, he quickly steps into the dirty, murky waters of the Bark River. Immediately the waters become a cool, sparkling circle radiating outward from his ankles, until the water is crystal clear, continuing upstream into Crooked Lake, and downstream in the Bark as far as one can see.

He utters a Hebrew blessing, "*Baruch Atah, Yehovah Eloheinu melech ha-olam.*" "Blessed are You, Yehovah our God, King of the universe." Then he cups his hands into the crystal water and offers a fresh drink to each of the young men. They all accept, sip the clean water, and immediately feel the refreshment on their lips, down their throats into their bodies. Aaron takes up an empty can nearby and washes it. They share more of the water with each other.

The Kadosh cautions the men. "Drink and refresh yourselves, but slowly. Let your dehydrated bodies adjust."

Isaac is first to finally get some words out from his now-moist lips and throat. "Are you the Messiah?"

"No, I am one of His servants. My name is Yoni. I am sent here to rescue you and many others in this region.

"I know that you have seen the Messiah's coming in the sky hologram-scroll and you have heard the trumpet sound. That was the sound of the last, the seventh, and the great trumpet. The hologram-scroll was projected from Jerusalem so there can no longer be any misunderstanding, no further confusion and no doubt about who is Messiah."

Ian, raised an Irish Catholic, is confused, concerned, and fears judgment or punishment. He asks, "Why're you here? We're fighters, Marines. We've killed a lot of folks. So what're you going to do with us? Did you come here to finish us off? Are you gonna throw us into a hell fire? You know, ya' gotta understand, all those guys we killed . . . they were all bad guys, really bad."

Yoni looks at Ian and pauses before replying. "No! No! No! Do not fear. I am here to save you, to bring you back to health and restore your strength."

Aaron asks, "And then what happens? Are you just saving us for some kind of judgment?"

"No, I am here to help you and others who have also survived these terrible wars, famine, plagues and worldwide catastrophes. Some other folks are near here, even now. I will help them too. And when all have revived, I am to gather you all for an adventure, for a return to the land of your fathers, the land of Abraham, Isaac, and Jacob."

"To Israel?" The overwhelmed four say almost in unison.

Aaron and Ian look at each other, both thinking the same thing. *We're not Hebrew. We are not Jews, maybe he's got us mixed up.* But they hold back their thoughts, deciding to keep them to themselves for the time being.

"Yes, I know what you are thinking," Yoni says slowly and deliberately, "but you are part of that remnant that is to be gathered, actually regathered, foretold in prophecies from over 2,500 years ago, by such prophets as Isaiah, and Micah, and Jeremiah, and many others.

"Isaac, can I show you all something in that old Bible your Aunt Ashley gave you?"

"Sure, I have actually been reading it some. But how did you know that?" Isaac hands over his stained Bible.

Yoni opens it and reads from Isaiah. Isaiah 11:11–12, "And it shall come to pass in that day, that Yehovah shall set his hand again, the second time, to recover the remnant of his people, which shall be left from Assyria, and from Egypt, and from Pathros, and from Cush, and from Elam, and from Shinar, and from Hamath, and from the coasts of the sea.

"And he shall set up a sign for the nations, and shall assemble the outcasts of Israel, and gather together the dispersed of Judah from the four corners of the earth."

Yoni hands the Bible back to Isaac. "Keep reading, you will find it becoming clearer, and very personal. Your lives are spared for a great purpose. This has always been in the plan of the Almighty One, blessed be He, long before you were born.

"I know that you all thought you may be the only people still alive. Now you know that there really are others. And I know you have hundreds of questions. But be patient. They will all be answered, in due course. There is plenty of time ahead, and much to learn.

"But first, let's get you cleaned up. Go ahead, jump into the newly cleaned stream, and scrub thoroughly. Here's some soap." He throws each a bar.

"I have prepared fresh garments for you and new footwear. Your Marine and camouflage garb has seen its day. Put these on after you bathe and are refreshed. Once you are dressed, you will find fresh bread and butter behind that log. So wash, clothe, eat and rest.

"I have to visit some others nearby. Shalom, *l'het-ra'ot.* Peace, see you later."

Yoni lifts off, slowly at first with his head up and arms held high. He gains speed and altitude and disappears toward the southwest. He is over the horizon within a few seconds.

Ellis says, "Well, now we know . . . he's a superhero."

The men do as instructed, cautiously peeling off their rotting clothes and easing into the clear, running water. It is cool, but very refreshing.

Ellis plunges under the crystal water, staying under for twenty seconds. As he surfaces, he shouts to the others. "This seems like a mikvah."

Aaron and Ian do not know what a mikvah is, but it looks like a good idea and they follow suit, moving quickly into the fresh, cool waters. The young men do not hurry to get out, enjoying their first bath in clean water in months.

Aaron is first to get out onto shore. As he comes out of the water, he notices the sores on his legs are gone and his dislocated left shoulder works full range, without pain. On a rock he finds four folded new shirts, pants, undergarments, socks, new hiking shoes, and a vest with short white and blue threads at each corner. Aaron is dressed, looking prim and proper, by the time the other men ascend from the cool stream.

As Isaac leaves the water he realizes there is no pain in his broken ankle. Warily, he puts increasing weight on it.

"No pain!" he yells.

Then he sees that the open wounds on his legs are now closed. Ian and Ellis too discover no aches, pains, or wounds. Coming out of the water, all four men are not only refreshed, but healed. Each man smiles at the others and simply nods. Ian, noticing the fully dressed Aaron, points to Aaron's new pants and shirt.

He asks, "Where'd you get those?"

Aaron simply looks to the pile left for each man. Each man slips into the new, clean clothes. Ian's new clothes fit precisely, as do the ones for Isaac and Ellis.

When clean and dressed the four walk over to a fallen tree remnant where fresh bread, twisted like challah bread, is set. And a fresh log of butter lies beside the bread.

Isaac is amazed , but pauses to remember. He goes for his Siddur, thinking this is a very appropriate time to give a blessing. He turns through several pages. "Here it is!" He holds the bread high with his right hand and repeats, "*Baruch atah, Yehovah, Eloheinu Malech Ha-olam. Ha-motzee lechem min ha-eretz.*" "Blessed art Thou, Yehovah our God. King of the universe, who brings forth bread from the earth."

Each one attempts to repeat this after him. He repeats it enough for even Aaron and Ian to get it down. Now the tune for the blessing is gradually coming back to Ellis, from childhood many, many years ago. Soon they are all singing the blessing with its tune over and over until they all know it well. Each man is also internally contemplating the profound and emotional meaning. They continue to sing the tune over and over again, until indeed, a rare event . . . four US Marines are crying, out loud.

Ian says, "Oh, the bread and butter and fresh water are SO GOOD!"

"Ooo rah!" They all repeat together, "Ooo rah!"

There is just enough to satisfy each man, without tempting them with overeating. After bathing, dressing, eating, and singing, there is silence as each one quietly sits on the ground and thinks long and hard about the events of the last few years, and especially of yesterday and today. They do not talk much. It is a time for reflection well into the afternoon. With their hunger satisfied and the months of accumulated fatigue surfacing, and not knowing what more to say at this point, they decide it is time for a nap.

Three hours later, as the autumn evening of Wisconsin begins to bring on a chill, all awaken nearly at the same time. Yoni is sitting on a log, across from a fire he has made. He seems in a jovial mood.

"Soup anyone?" he asks. He has prepared a three-liter pot of soup, held by a wooden fork frame, suspended over crackling oak logs in a new stone-ringed fire pit.

Ian blows on the soup and downs a few sips. "Man, is this ever good soup! Where in the world did ya get all these ingredients, even spices and the Wisconsin cheddar cheese?"

Aaron adds, "And the spoons and bowls?"

Yoni winks and smiles. "My secret, for now."

As the shadows of the evening are lengthening, Yoni and the young men find themselves in deep discussions and, finally, a lengthy question-and-answer time. Yoni is warm and sharing, however, his answers are frequently evasive, promising more information when the time is right. More than once, the four were reminded a new world order was in place and answers will come, but until then . . . be patient. There is so much to learn, and it is becoming increasingly apparent, so much to be done. Yoni is surely a lifesaver. His miracles of clean water, clothes, food, comfort, and healing are much appreciated.

Yoni says, "But what I am telling you, is that life going forward will not be just miracles. It will not simply be life handed to you on a platter. You and others have been saved to assist in a great work. We all are privileged to assist Yeshua, the Messiah, in the rebuilding of the planet. As you will come to understand, you all will greatly benefit personally from this new world, as will so many, many others. And better yet, we all will have a hand in its reconstruction. Yes, a few miracles every once in a while may be needed, but the major rebuilding will be sweat, tears, ingenuity, determination, and old-fashioned hard work."

The discussion lasts well into the evening. Then Yoni gracefully lifts off into the southwestern night sky. They all have talked out, talked to exhaustion. Their throats are getting dry. All gradually fade.

Isaac says, "I think it's time for another long night's sleep. There's a lot 'a work to be done in the morning."

DAY 3 WISCONSIN

Everyone awakens early, just as the first rays of the morning sun begin to diffuse through the still rusty, thick atmosphere. Yoni has not returned. Ellis is first for a cool fresh bath in the crystal clear Bark River. He jumps in, cannonball fashion, making a big splash and with a loud "Ooo rah."

Aaron, Ian and Isaac respond with their own "Ooo rah" and all join Ellis. Three more cannonballs! The clean, cool water is so refreshing, it is hard to get out, but the morning air is a bit chilly. They had for a moment forgotten that this is mid-autumn in Wisconsin.

"Where are the towels? He forgot the towels! What kinda service is that?" Ellis jokes.

The other three don't laugh or even smile, somewhat wary of any critique of this still unknown superman.

The gall that Ellis has, Isaac thinks.

Afterward, clean and freshly dressed, Isaac finds and gathers some low-growing raspberries around the back of the old building where they have been seeking shelter. Why had he not seen them before? Isaac rejoins the others and again digs out his Siddur, turns a few pages and finds a blessing for fruit. He recites the Hebrew blessing for them, before devouring the berries. "*Baruch atah, Yehovah, Eloheinu malach Ha-olam, borei prei ha-gafen.*" "Blessed art Thou, Yehovah, our God, King of the universe, who creates the fruit of the vine." The other three only listen, then all dig in.

Looking up in the distance the men can see a now familiar bright light coming their way, and they watch with anticipation, knowing that the light is the return of Yoni over the horizon.

"*Boker tov!*" "Good morning," Yoni remarks as he touches ground in front of the log where they are sitting. "I see you found the berries behind the building. Good.

"I have located several more groups of remnants in this region. Relax. They too are now well and safe from further harm, but now we must plan a rendezvous and get on with our travel plans."

Isaac asks, "Travel plans? Travel plans? Uh' so … so … so what's the plan?"

Yoni smiles, takes a pause, and again recounts the prophecy of Isaiah 11. He explains, "Now, at this moment, all over the planet, Yeshua is directing not only me, but my brother and sister Kadoshim to gather the remnant of Israel back to the land of our Fathers, of Abraham, of Isaac, of Jacob while helping the millions of the nations scattered around the world."

"Millions?" Ellis remarks. "You mean millions of people survived these wars, catastrophes and disasters?"

Yoni nods, but does not provide any more detail. "Our primary goal is to return to that land and rebuild it. From there on, we will guide and lead in retraining the world and rebuilding the whole planet. That's the big picture. But, today we are starting out with much smaller goals. Everyone on board?"

"Ooo rah!" the four shout.

"Now, here is the immediate plan. I want you to take the tools I have brought you to build a raft that can carry fifteen people. After building it, you will punt down the Bark River. Along the way and where the Bark flows into the Rock River, you will meet up with others. From there your larger group will travel down the Rock River and eventually the Mississippi. Along the way and at each junction I will have directed others to join you.

"Once at the Mississippi, others will meet and then together . . . on to Israel."

Ellis leans forward, pauses, then observes, "Wow, that's a bit of a plan! Guess we'd better get our butts in gear and get busy. Where're those tools?"

Yoni pulls an assortment of tools from his white bag, empties them on the log. "Yes, I too had better get going. Don't want to hinder you from your work." He smiles and winks. Then he is off, disappearing over the western horizon. He's off before Ian or Aaron have asked their lingering question. Why Israel? Israel, for them, at least?

Isaac walks away to the northeast and is gone for twenty to thirty minutes as he surveys the situation. Returning to the group, he reports his assessment. "The Bark River is really too narrow for a wide raft in these parts. But we need a raft to eventually hold fifteen people, as Yoni's ordered. So some design compromise on the raft has to be made. We can make it a bit narrower and offset that by being a bit longer. Even then, where we build it and where to put it in is also critical."

Aaron explains, "When the Bark River comes out of Lower Nemahbin Lake, it's more fit for a canoe or kayak. It takes so many tight hairpin turns and the sandbars and shallows are frequent. Since we're located just below where the river enters and then quickly exits Crooked Lake, down-stream's where the river becomes a bit wider and deeper.

Isaac nods his head. "Yeah, where the water moves more quickly as it heads south toward Dousman, is where we should build the baby, and put it in."

All are in agreement. They follow the river downstream for a quarter-mile. Ellis gets some paper from the back of Isaac's journal and sketches a plan. The men hurriedly scurry about the surrounding area, finding limbs, vines, identifying fallen and burned trees and twisting small vines into ropes. The pieces for the raft are laid out on the sand bank. The vine-rope is woven between the parallel logs and an overhead cover is built toward the front of the raft.

With the tools left by Yoni, along with their carefully trained survival skills, perfected in their long flight from the Gia Union Troopers, they manage to assemble the rustic raft by just before nightfall. Each man feels like a teenage Eagle Scout earning his final merit badge.

Ian looks very self-satisfied. "Now, that's one helluva good-looking raft!"

Ellis carves out four long poles to push on the bottom of the river, that is, to "punt" the craft along the shallow river. Isaac tries to fit a flat wooden rudder onto a fifth pole, but it doesn't seem to feel efficient. Punting will probably have to be the power and the means of control.

They are ready and anxious to get underway, but spend the rest of the oncoming evening darkness gathering new-found berries and roots for food along the way. They put their tired and scorched sleeping mats and their new supplies on the raft where they will sleep the night, with one-man rotation, two hour watches, still distrustful and on the lookout for Gia Troopers that could attack.

DAY 4 WISCONSIN

Morning breaks. A new day, a new adventure!

Should they push off and start downriver, or await directions of Yoni?

Isaac says, "Yoni seems to be really busy and he has already instructed us. We know the Bark River like the back of our hands. I'm stoked. So, let's go!"

All agree. They slide the raft into the narrow stream, a quarter-mile beyond the Crooked Lake outlet. Each takes a corner with his punting pole. Each pushes in unison and the raft begins to drift with the small slow river current. Isaac strains trying to maneuver his tiller to keep his rudder at a slight angle, as it is asymmetrical. However, as he had feared, the rudder finally breaks from his pole.

"Dammit, we'll have to navigate with putting poles only," Isaac says loudly.

After a while Ian becomes discouraged. "Hell, I think we'd make better progress walking over ground, rather than this slow-assed rafting down the Bark."

Everyone pauses, but after a while, Ellis also expresses this same feeling.

Then, after some time, Ian has a change of heart. "I'm just thinking and guess I should remind us . . . this land has been parched and burned and there are fallen trees, rocks, burned out vehicles and all kind of crap blocking the roads and trails. That's something we exploited in avoiding the Gia Union Troopers, who were hot on our tails, but it sure as hell won't be an advantage now."

The reminder is sufficient to give the group comfort that the decision to raft the river is the correct one. The raft is just low enough to go under the bridge of Summit Drive and then the Highway 18 bridge, but just barely. The last bridge under Highway 18 is the toughest as the river narrows due to a landslide under the bridge. But they are able to maneuver the raft fifty degrees starboard then forty degrees to the port, wiggling thru the narrows.

They are beyond the bridges when Ian spots an ominous structure ahead. "Oh crap, guys! Look! Over in the distance on the right!"

They can now see the remains of a Gia Union Command Center in Dousman.

Even though Ellis knows world government has changed, he still has guarded respect for what has once, and so recently, been so much feared. He whispers, "Imagine, we were this close to a damned Gia Union Command Center! Good thing we didn't know it. Would have never got any sleep. Think about it! Each friggin' center is a station for five thousand Gia Union Troopers, and they were just six miles from us!"

But there are no Troopers to be seen today and no vehicles. In fact, there is no sign of any life around the Command Center. The same in the town of Dousman, as they float past the town center and the once popular Bark River Park on the western edge of town.

They move south and west. Tributaries into the Bark are now adding to its volume and depth. It gradually flows faster. Soon there are fewer sandbars and shallows. Still, fallen trees and debris frequently hinder their voyage. Sometimes they can raft around it; sometimes they have to stop and clear the channel. Along the way, the water remains clear and clean, so they now dare, without any ill effect, to drink directly from the river with their cupped hands.

Ian lies on his back for a rest and looks upward. "Maybe I'm just getting used to this rusty crappy air, but it seems to me the sky is getting clearer. I think I can even see some blue now and maybe even a whiff of a white cloud. But maybe, maybe it's just my imagination."

Aaron, Ellis, and Isaac look upward and stare, and all nod their agreement.

Isaac casually turns toward Ellis. "This all seems so much like a dream. Somebody should pinch me . . . make sure I'm awake."

Ellis of course accommodates and gives him a sharp pinch on his arm.

Isaac winces. "Thanks, you obliging bastard. Yeah, I'm awake alright."

Ellis raises his eyebrows and exaggerates a cocked smile. "Yeah, but . . . maybe the pinch sensation is part of your dream? Um. . .? Du-da-du-da-du... Me thinks maybe it's The Twilight Zone."

The sun has now passed high noon and Aaron takes charge. "We should pull over for a rest and a pit stop."

But the other three are all still a bit edgy about Gia Troopers in the area and tell Aaron to do as they have and make do with the overboard facilities. "Just pee in the river, and keep punting," Ian says

Isaac after a while responds, "Well still, I think Aaron's got a point. We all could use some rest, so let's look for a high bank to minimize our exposure."

For the next twenty minutes they drift and look, then Isaac spots a place. "See those granite cliffs coming up on the starboard side? That'll be a damn good location. Did you guys know those cliffs were carved out by glaciers during the last Ice Age? The Glacial Drumlin Trail parallels the Bark River here for several miles, on the north side."

Ian yells, "Look, there's a flat shoreline between those two cliffs. It's along the trail, on the right. And there's some burned out tree stumps to tie the raft."

"That's the place," Isaac says. "Ellis, secure the bow line. Aaron and Ian, get the stern lines."

Smoothly, alongside they come. They cautiously survey the shoreline. Aaron climbs up to the cliff's summit and peers into the distance sides of the river, motioning 'all clear.' All climb to shore, stretch out and then sit or lie down, with one designated as rotating watch.

After a break, as they perch on the rocky shore under the shade of the outcrop, Isaac feels like leading the team in a brief reflection of their transformed state and to giving a thanksgiving prayer to the Almighty One and his Messiah, who have just brought such dramatic change and optimism to them, at a time when there was no more hope and death seemed certain.

He begins, reflecting what he has been reading from his Bible. "Blessed are you, God of Abraham, God of Isaac and God of Jacob. Our Father, we pause to thank you for lifting us up out of the net, when the snare of the trapper was certain upon us. You alone looked down upon us, and have given us life, hope and a future we could not, and even now do not, fully imagine."

The others look at each other, nod, then together say "Amen."

They each take a few bites of the berries and roots they have stowed onboard, along with a couple bites of the bread given them by Yoni. They eat and rest.

Aaron gazes downstream. "No sign of Yoni yet today. Perhaps we should've waited for his command to proceed?"

Ian says, "No, I think, from what he told us, he'd expect us to take initiative, do our part. And if we screw up, he knows where we are and can find us … even come to our rescue again, if necessary."

Rested, they all concur as they push the raft off the rocky shore heading downstream toward the convergence with a much bigger and wider Rock River.

Ian observes, "Guys, we're traveling at a quickening pace. The river is getting deeper and not as many of those friggin' blockages. I think there's a faster current speed cause of the increased volume of water."

Even so, as the first day's light begins to fade, they have traveled only twenty-five miles and, due to the meandering course of the stream, only twelve or so "as the crow flies." Isaac thinks the trip to the Rock River junction near Fort Atkinson could require another two or three days. He is uncertain, and wants input.

Isaac asks, "Should we pull over for the night or raft on thru the night? What do you all think?"

There is no question about it. The four push the raft to the port shore, onto a sandbank, which is littered with broken trees, driftwood and debris. Ellis gathers logs from the fallen trees and sets up a makeshift table from a piece of driftwood. Aaron has fire tender and a flint, so a campfire blazes soon on the bank. Ian and Isaac find reeds from a marshy area adjacent to the sand bank and harvest as much as they can carry. Back on the bank, they weave and tie them into sleeping mats.

"We can use these for sleeping mats, as well as pads on the rafts. And we may need additional sleeping mats, if we really do pick up more passengers," Isaac says.

Still leery of Gia Troopers, it is decided to keep a perimeter watch. The four will take two-hour shifts. Everyone feels they can sleep better with that precaution.

Day four ends. It is night one on the river.

CHAPTER 4

WHERE'S GOG?

DAY 4 EVENING: SCUPPERNONG RIVER

Yoni has located other survivors in desperate need of his rescue. Right now he revisits a small band of five, located on the east bank of the Scuppernong River, not too far away from the four Marines in southern Wisconsin. Sitting around a campfire after sharing bread, butter, berries and melon, he asks if they would like brief report of what is going on in and around Jerusalem since the dramatic arrival of Yeshua.

As students at Whitewater University they had kept up with a lot of what was happening, but their information was fragmented and frequently based on rumor and hearsay. They, also, more recently had been on the run and had lost contact with other survivors.

One member of the small group, Teah, glances at her four companions Michael, Rivka, Raquel and Miriam. Like her, they are eager to get some factual first-hand information.

Teah says, "Yes, please tell us as much as you can. But, right now, what I don't understand is . . . where's Gog and all his armies? We saw the hologram images of Jerusalem and Yeshua's entry, but I can't figure it. It seems strange that Yeshua and the Kadoshim found Jerusalem empty, with no resistance."

Yoni understands their need for more information. He says, "Well, we have the evening before us and I'm ready, even anxious, to give you as much of an update as you want. I just don't want to pile too much on you at once or to bore you."

Raquel says, "You must have forgotten. We're graduate students. Due to excessive exposure, we are immune to information overload and boredom."

The group chuckles and nods their agreement.

Yoni reassesses his plan and decides they are entitled to more descriptive detail and also would undoubtedly be pleased with some historical references.

He smiles and winks. "You asked for it."

Yoni begins …

Well, actually it's not accurate to say there was no resistance. But there was very, very little. There was minor opposition from a surprised Olivet Security Unit of Gia Union Troopers who were quickly brushed aside with a blast of His voice. Two other small Security Units, one on Mount Herzl and another near Talpiot, scattered and fled. The Mount Herzl unit spread out to the northwest, toward Modi'in while the Talpiot unit headed south toward Tekoa.

Remember, to provide protection, the city of Jerusalem was built where it is today due to its high steep hills, deep valleys and potential for high security walls. But now, in the age we live in, it is essentially unprotected because even those magnificent historical walls around Jerusalem have no defensive purpose in an era of rockets, cruise missiles, laser and particle beams and of course, airborne assaults. However, in 1540 AD, when those walls were built at the instruction of the Ottoman Sultan, Suleiman the Magnificent, they really were important for the city's defense.

Indeed, as you know, Gog, the Supreme Commander of Gia Union Forces, does not rely on those walls either. Rather, he relies on a worldwide system of modern integrated air, land and naval armed forces, supported by extremely advanced technology. His military structure is based on 'offense,' with his aggressive system arranged at tactical locations, such as the fifteen-hundred powerful Gia Union Command Centers, strategically placed, around the world, all tightly linked by satellite communications.

Normally, around five-thousand Gia Troopers are stationed at each Command Center and, if necessary to suppress any resistance, additional Troopers can be deployed at a moment's notice. In fact, I would guess that none of you know that near here there is a Center over in Dousman and another in Fort Atkinson.

For Gog's system, rapid deployment is essential. Sea-based Troop carriers, called Gia Cruisers, can carry one thousand Gia Troopers each. They are hydro-foils and can speed the troops at forty knots, with an unlimited range. And through the air, massive Air Cruisers carry up to four hundred troops, along with tanks, artillery, armed personnel carriers, weapons and supplies.

Thousands of spying Sky Pod Intellites hover continually over most major cities and over other strategic locations, providing instant visual surveillance. The Sky Pod Intellites are also equipped with remotely activated Superfire X missiles.

Gog has destroyed so much and so many lives. But we knew he was coming. He has been referred to by many names. Believers in Yeshua and Jesus have dreaded his prophesied coming for thousands of years. The Bible refers to him as 'the man of sin,' 'the Beast,' 'the Evil One,' and 'the Antichrist.' In the ancient Hebrew prophecy of Ezekiel, he is known simply as Gog.

He has even been given a mysterious, kabalistic number to his name, 666. And now he is here, and has been here for many years.

The book of Revelation predicted him to be working in concert with a 'False Prophet,' and would be directly controlled by HaSatan (Satan, Lucifer) himself. He has adopted the title HaNavi Rav, which means the great prophet, but his name is Dajjal, meaning deception in Arabic. These three personalities have been described as "the unholy trinity." They were prophesied to work together at the end of the age and to finally achieve, at long last, a one-world government. That, they did achieve . . . for a while.

You know, the name for their one world government is especially interesting. It was chosen to be the Gia Union because Gia is a shortened version adopted from Gaia, the name of the Greek goddess of the earth, often for centuries just called "mother earth."

From his headquarters in Jerusalem, the Gia Union Supreme Commander has been able to marshal whatever forces are required to control any situation. Over the last few years, the Gia Union forces have been able to systematically tighten Gog's control over "mother earth." You know much of it, but probably not from the Jerusalem perspective, so I shall continue.

For a moment, let's go back in time, some years in time, as seen from Jerusalem. Once Gog was appointed Supreme Commander of the Gia Union, he made high priority of confirming a treaty with Israel. This treaty made Jerusalem an international city and the Headquarters of the Gia Union. Jerusalem would become the capital of the world. For Israel, there would be worldwide freedom of commerce with no restrictions and complete Israeli rule from the Jordan to the Mediterranean. There would be no divisive Palestinian state.

Most important of all, there would be guaranteed freedom of worship. Muslims, Christians and Jews would have sites of worship on the Temple Mount itself. The Gia Union would guarantee those freedoms. The Jewish place of worship would be a small fabric-covered Tabernacle set up on the site of the former Dome of the Rock, which had been destroyed in a recent earthquake. No, there would be no Temple, but would be a fully functioning Tabernacle, on a site the rabbis felt the holy of holies had been.

The Muslims would continue worshipping in the repaired and restored Al Aqsa Mosque, on the south side. The Christians would have an ecumenical church north of the Tabernacle to be called The Nea Church II of Mary Incarnate. This was a reconstruction in memory of the massive Nea church built by Justinian in the 6th century, and destroyed by invading Muslim armies shortly afterward. From those ruins of the original Nea

Church, over on Mount Zion, many of the implements of the Herodian Temple had recently been recovered. But that's another story.

And so, Gog set up operations in Jerusalem. A central Gia Union Command Center was set on Jerusalem's highest hill just to the west on Mount Zion, overlooking the Temple Mount. The site was located over the archeological site where Herod's palace had been two thousand years before. The archeology was preserved in the basement of the Command Center.

With this and other contracts, the whole world had become integrated into one economy, controlled from Jerusalem by the Gia Union. Commerce and the economy were tightly managed through data received from The Marck. This ensured an accurate and centralized database for all activities, and for all people. A worldwide, efficient central government control system had finally been achieved. With this new progressive government, there was a never-before experienced optimism for the future. It looked like a great new wonderful world.

Instantly there was an enormous increase in global efficiency. With a reduction in waste and inefficiency, there was a corresponding burst in economic activity. Increased investments, coordination of activities and worldwide major building projects ensued. The next three years were a time of unparalleled prosperity beyond anyone's expectation. It seemed as finally mankind had achieved a wonderful, wonderful, new world.

Great advances ensued in science and medicine. These were accompanied by the improved health and welfare of the earth's population. Physical and social wellbeing were obvious. Urban projects improved housing for everyone. Construction of new homes, businesses, upgrades to infrastructure, as well as the building of the new Gia Union Command Centers, with high technology weapons and supplies, all added to the general prosperity.

The Gia Union gained control of all national, regional, local and individual economies and activities. Well, almost all. There were, surprisingly, millions of individuals who refused to accept The Marck. The Marck was a zero tolerance issue. Most people complied, but ultimately thousands had to be terminated. If one didn't allow the implantation in their right hand, their right hand was cut off. Then they were offered implantation in their forehead. If they refused, then off went their heads. All was done in a public forum. Gog directed that an aggressive hunt commence for the dwindling numbers who were still defiant. Defiant folks, well, like your brave group.

Then the big event, the REALLY BIG event, happened. The Supreme Commander perceived that he had reached the pinnacle of worldwide power and that there was no effective opposition. Gog reasoned, 'My plan is now in full effect. It has taken decades to develop, but over the last three-and-a-half years, I have succeeded. I have achieved absolute rule. Now, I have reached the zenith of the Gia Union success. It is now time for me, the Supreme Commander, to confirm and finalize my Divine, God-like, status.'

The Supreme Commander then arranged a worldwide telecast from the Jewish Tabernacle on the Temple Mount. A universal holiday was proclaimed that would coincide with Passover and Easter week and the middle of Ramadan. All three holyday seasons coincided that year.

All worldwide communication networks were live and the drama was broadcast globally. By Gia Union orders, all types of media were to carry this real-time broadcast . . . and only it.

Trumpets were sounding, bands were playing and there was dancing in the streets. Great light and sound displays served to highlight the Tabernacle while great elaborate rituals continued in the courtyard. A great levitical choir performed, with orchestral accompaniment, while the traditional sacrifices of lambs, rams and bullocks continued throughout the midday.

When it was time for the sacrifice of the Passover lamb at three in the afternoon, Gog stepped into the inner courtyard and proceeded to the bronze altar. He presented his sacrificial Passover lamb and then stepped forward to personally slit its throat. To everyone's surprise, he personally collected its blood and with ritual and drama passed it to the High Priest. The High Priest sprinkled the blood on the four horns of the altar, then skinned the lamb and placed its body on the roasting fire.

The Supreme Commander then ascended into the Holy Place. Without any obvious respect, he walked proudly through the Holy Place, passing by the golden menorah, the golden face-bread table and then the golden alter. He strode right up to the inner curtain. There he momentarily paused and took a deep breath. Then he dramatically and personally opened the double layer of the inner curtains that separated the Holy Place from the Holy of Holies. He proudly presented for the world to see, the inner sanctum of the Holy of Holies.

Gog's two trusted body guards then held the curtains open so all could see him as he walked westward into the Holy of Holies. Then turning fully one-hundred-eighty degrees to face the video equipment to the east. With his back to the Ark and the Mercy Seat, he said, 'No one in history but me has achieved so much, so fast and so effectively. No one has done so much for so many as me. Worldwide prosperity and unity have been obtained -- by me. So let it be known to all, I am the Savior of mankind. Yes, *I AM YEHOVAH GOD!*

'I am the One God, the Almighty, the Alpha and Omega, the Aleph and Tav, the King of kings, the Lord of lords and Savior of the world. I will now accept your worship.'

Still with his back to the ark and the over-shadowing cherubim, he walked eastward back through the opened double curtains. A personalized, custom designed crystal and gold throne was brought into the Holy Place and placed just to the east of the golden alter, facing west. Gog paused to admire his throne, then suddenly and dramatically sat down on its opulent and lavish seat. The throne was situated equidistant between the golden menorah on his right and the table holding the face-bread on his left, facing east. His back was deliberately to the golden incense altar, and behind that, the Holy of Holies with the Ark of the Covenant.

He sat down and smiled broadly for a long time, calmly waving to the media center recording the event. He was thoroughly enjoying this long planned event. Then, as the first wave of cheers subsided, his Prime Minister, Dijjal (the False Prophet) came in from behind the menorah and paused. The choir and orchestra approached its crescendo and, with great drama, he placed the golden custom crown, a golden and crystal crown with ten horns, onto Gog's head. Gog was now, for sure, officially the King of kings, and Lord of lords. All the world saw.

The music and celebration reached such a new crescendo that the sound rattled and shook windows throughout Jerusalem and the hills.

A week of worldwide celebration and festivities had been proclaimed. The celebration was not limited to Jerusalem and festivities were ongoing in the world's major capitals. The broadcast from Jerusalem was also on the big screens in Times Square, Chicago, Los Angeles, Paris, Rome, Berlin, Tokyo, Beijing, New Deli, Istanbul, Sidney, all the other major cities and even in smaller towns and villages. After Gog proclaimed he was the real God and Savior of mankind, millions of people cheered and danced in the streets.

From his throne, Gog was served the roasted flesh of his Passover lamb. With disrespect and his back to the Holy of Holies and the ark of the covenant, he ate his Passover lamb right there, complete with unleavened bread and wine.

This event was not unexpected. It was prophesied by the prophet Daniel and then quoted four hundred years later by

Yeshua himself. We read in the book of Matthew from two thousand years ago:

When you therefore shall see the abomination of desolation, spoken of by Daniel the prophet, stand in the holy place . . . then let them that be in Judaea flee to the mountains . . . For then shall be great tribulation, such as was not since the beginning of the world to this time, no, nor ever shall be. And except those days be shortened, there shall be no flesh be saved alive, but for the elect's sake, those days shall be shortened. Immediately after the tribulation of those days shall the sun be darkened, the moon shall not give her light, the stars shall fall from heaven, and the powers of the heavens shall be shaken (Matt. 24:15–29).

And, sure to this prophecy, Gog's proclamation and desecration of the Holy Place coincided with the beginning of the Seal-Plagues, the Trumpet-Plagues and the Bowl-Plagues. Also there was growing resistance to his government. It started right in Jerusalem.

To his surprise, instead of understanding and appreciating his proclamation, Jewish and Israeli leaders sent a strong letter of protest to the Supreme Commander, which he of course ignored. And as he continued to broadcast events of official business from his throne in the Holy Place, those same Jewish authorities and Israeli government officials became more belligerent. They eventually organized protests in the streets of Jerusalem and around the country. Then the street protests became more vocal, and naturally degenerated into riots. These protests became daily occurrences in Jerusalem, as well as in Tel Aviv, Modi'in and all over Israel.

Finally, one day as the protesters approached the Temple Mount, Gia Union Troopers opened fire, first on the leaders. The protesters kept coming and the Troopers kept firing. Over one hundred were killed. Over the next days, bigger protests ensued and the violence worsened.

Since the protesters all had The Marck, the Gia Union system knew who they were and where they lived. The second night, Troopers spread out all over Jerusalem arresting the protesters along with their families. This further inflamed the population. The Supreme Commander then ordered executions of over two thousand protesters along with their families. Over ten thousand executions were publically broadcast. Still the resistance continued and thousands more were rounded up and killed. Only then did many of the population heed Yeshua's two thousand--year-old advice and flee the city into the mountains and desert areas. Over the next few weeks, Jerusalem was depopulated with only Gia Union officials and Troopers remaining. Over the next few months the same situation continued in Tel Aviv, Modi'in, Netanya and Beer Sheva.

Inevitably a hundred Israelis, former Golani Brigade members, organized an assassination attack on Gog himself. They referred to themselves as the New Irgun. They planned a storming of the Tabernacle during one of Gog's personal appearances in the Holy Place.

But Gog had been tipped off by his prime minister and spokesman, Dajjal, who had received direct intelligence about the plan. The Israeli troops did not appreciate the extent of the Gia Union intelligence systems or the capability of their technology. They had no idea it was a trap. They met minimal resistance as they made their way from the Pool of Siloam through an ancient tunnel leading directly to the underground cisterns below the Mount.

As they entered the Tabernacle's east court curtains, they noticed the video equipment and lights focused on the menorah and alters. But Gog was gone! The Israeli soldiers suddenly found themselves surrounded by over a thousand

Gia Special Ops Troopers. With video systems active, Gog ordered the fire-bombing of the whole tabernacle structure with the Israeli soldiers inside. The Commander had the scene broadcast for the world to see. And the world concluded, once more: It was the Jews' fault.

Well, this Tabernacle had endured only 42 months, three and a half years. And its destruction occurred on Tisha B'av, the 9th of Av, on the Hebrew calendar. That was the same calendar date, Tisha B'av, that the first Temple was destroyed by the Babylonians in 586 BCE. And again, the same Hebrew calendar date of the destruction of the last Temple by the Romans in AD 70. It's interesting , not a lot of people know that it's also the same date as the expulsion of the Jews from Spain in 1492.

Gog then immediately issued orders for the confiscation or destruction off all Jewish property within Israel. With worldwide public support, Gia Trooper swept through the cities, villages, suburbs, industries, farms and orchards all over Israel. Jews caught within the land were immediately expelled or executed. Then at the suggestion of Dijjal, the Gia Union employed a policy of population replacement. Arabs, Turks and Egyptians were invited to 'come back home,' with free land, houses and no taxes for ten years.

Let's fast-forward to now, after three years of plagues and earth-shattering events, Gog was losing control. He has had other regions rebel, not just the Israelis. His worldwide reign of prosperity and unity has been unraveled. Yet, he still believes he can depend upon the Gia Union Troopers to restore order.

Only four weeks ago, there were indicators of another regional rebellion to the north and east, in southern Russia and Ukraine. Communications with the Gia Union Centers in those sections had been disrupted and no tax revenue had been coming from those territories.

Gog publicly speculated how it could be that the federal states in those regions were in rebellion. 'I severely crushed three other rebellions last year. These new rebels need to be taught an especially powerful lesson, a severe punishment, one for everyone to see and understand. I'll lead the armies myself to personally crush their revolt. I've had enough of these insurrections. Don't those people know when they have it so good? Don't they realize how much I have done for them? This kind of civil disobedience must be crushed, swiftly. I'll personally see to it,' he screamed to the media systems. The crowds cheered, 'Hoorah,' and chanted, 'Long live Gog. Long live Gog. Long live Gog.'

So, three weeks ago he began to assemble all the Gia Union Troopers from Israel, Arab League, India, Iran, China and the European Union. They focused on a staging ground in northern Israel, in the Jezreel Valley. Initial forces exceeded four million, which he commanded from the Megiddo Gia Union Command Center. More Troopers and equipment were pouring in when . . . when . . .

When, It Was Day One.

The mighty sound of the shofar and the appearance of the global hologram-scroll were witnessed all over the world, Har Megiddo was included. Gog and his four-million-man army in the Jezreel Valley saw and heard it at the same time as you and everyone else. With disbelief and rage, Gog personally watched Yeshua land on the Mount of Olives and then enter Jerusalem. Fury consumed him.

Suddenly, no longer was Gog's primary concern the rebellion in the north and east. He quickly conferred with his counselors, but the truth was apparent. His capitol, his center of operations had been captured! He had foolishly left no defense.

Typical of his conceited, maniacal behavior he thundered, 'How could Yeshua have been so deceptive? How could he sneak in when we were away?'

The Supreme Commander knew what was at stake and he knew with whom he was battling. He had expected this confrontation and was confident he would win. He just wasn't expecting it at this time, while he was away. How could Yeshua know when to come in with such a surprise, at a time when he and his forces were absent? What a trick!

There was only one thing to do, Gog reasoned, 'This is life or death, victory or defeat. All depends upon immediate action. The rebellion in the north will have to wait. In fact, maybe this is an opportunity to regain their alliance, for the common good. They have the same plan, are part of the same program.'

Urgent and emergency commands were universally delivered by whatever communication systems were still available. All Gia Union Forces, worldwide, were commanded to assemble immediately in the Jezreel Valley for a massive attack on the invaders in Jerusalem.

Gog made a personal appeal to the northern rebels to join in the retaking of the world capital and the ouster of Yeshua's new government. He realized he needed all his resources for it to be an overwhelming victory. He even rejected the principle of always maintaining a rear guard for protection. All Troops and Air and Naval forces from every Command Center must be assembled to move forward together in the attack on Yeshua and Jerusalem.

Gia Union forces from the north, south, east and west immediately accelerated their arrangements to join the assembly. Even the rebellious states in the north and east quickly recognized the urgent need to join in the attack. This is to be all-out-war!

For days, massive forces continue to come together. Over the next few days more, additional hordes of straggling Gia Troopers manage to arrive, even though they have been constrained by the global turmoil of the recent plagues. Many more were still on their way. The largest force in history was being marshaled in the plain adjacent to Har Megiddo, the hill of Megiddo, at the Megiddo Gia Union Command Center.

Air Forces and Air Commands are stationed in the surrounding countries of Jordan, Egypt, Lebanon, Syria, Turkey, Iraq and Arabia. Gia Navy Command have brought aircraft carriers, missile cruisers and more troop ships along the entire Israel coast and into the Red Sea.

You should know that as we sit here, right now, enjoying our meal by the warmth of our campfire, Gog's armed forces are rapidly massing for an all-out attack on Jerusalem. Their support and supply lines extend for hundreds of miles in every direction. The total force may exceed eight million, equipped with the world's most sophisticated weapons. These include chemical, biological and nuclear weapons of mass destruction. From the Har Megiddo Gia Union Command Center, Gog plans an overwhelming strike south upon Jerusalem, soon.

The group of five has been riveted to Yoni's report.

Teah quietly and humbly says, "Thank you so much. You've rescued us and fed us and we're no longer in the dark, not knowing what's going on."

Michael asks, "If the Gia forces are congregating at Har Megiddo and Yeshua is in Jerusalem, where's Gog personally?"

Yoni looks over the crackling fire and opens his eyes wide. "He is right now at the Har Megiddo Gia Union Command Center with his troopers. All that Gog has to offer points south."

Rivka asks, "What is Har Megiddo?"

"Har Megiddo is the name of a hill in the central valley of Israel, overlooking a broad agricultural plain stretching from the Mediterranean Sea to the Jordan River. *Har* in Hebrew means "mountain." It is only 60 miles north of Jerusalem. This mountain is, however, only a tel, and not a natural hill. A *tel* is an artificial mountain made of the successive remains of settlements, new ones on top of the older, over thousands of years.

"Multiple battles have been waged at Har Megiddo over the last four thousand years. It is famous for many of the decisive battles of antiquity as well as in more modern times. Here Pharaoh Thutmose III defeated a Canaanite alliance in 1479 BCE, and in 609 BCE Pharaoh Necho II battled the Babylonians. Good King Josiah of Judah was killed in that battle.

"In 1918, General Allenby defeated the Ottoman armies there and liberated the land from the Turks.

"And it is thus fitting that the prophet John forecast it to be the gathering place for this end-time, battle of all battles, really this time: the mother of all battles. Revelation 16:16 says, 'And he gathered them together into a place called in the Hebrew tongue Armageddon (Har Megiddo).' Har Megiddo is not prophesied to be the site of the decisive battle, but the gathering place for the forces engaging the Messiah, in the end-time final battle."

Raquel is quiet, pausing to digest and to reflect on all of this before speaking. "So, the battle is really about to start."

Yoni says, "Yes. Yes. Yes. It is. With a moment's notice, responding to the slightest signal from Gog, the Gia Union Supreme Commander, they will all move toward Jerusalem."

Miriam says, "Oh my! Thank you for all you have done for us. We feel so much better, but with the situation in Israel the way it is, and what's ahead . . . I don't . . .I just don't think we're going to sleep very well tonight."

The four others agree. The campfire burns, ever more dimly.

CHAPTER 5

OTHERS?

WISCONSIN - DAY 5

Morning comes with the first bright rays of the sun.

Isaac stretches and yawns. "I think I'm getting used to a bright sunrise again. It's been a long time, but now it's beginning to wake me like it used to."

Aaron says, "Well, I've been up for a while, checking things out. There seems to be no Gia Troopers. In fact, there are still no signs of any other life, man or beast."

The team eats a little, completes preparations, and pushes off early, to make the most of a full day.

As the group drifts further down the Bark River, they survey all the destruction, from the shoreline to the horizon. The raft often drifts quietly with long periods of silence. Each man is meditating, thinking about what has transpired over the last few years, the last few days, and what will lie ahead. Although Yoni has assured them that they are not alone, and that they will meet up with others, each has lingering doubts.

"I was just thinking," Ellis says in a low voice. "Considering how much the world has been through the last few months, and how close we came to dying, I'm not sure how anyone could be alive. Y'know, we're trained survivalists and we've barely made it. How could millions of people have made it too?"

There is no response from the others. Ellis is just expressing what everyone else is thinking, but not wanting to say aloud.

As the Bark continues westward, it enters a flat plain where the waters become shallower and the river widens. This is where, in the past, a dam and pond had existed for over a century. The area was settled by Cyrus Cushman in 1840, and has been in his family since. Early on, the Cushman family had built the dam to power a sawmill, which operated into the mid-twentieth century. The Cushman Dam was condemned by the Wisconsin Department of Natural Resources years ago, and then destroyed in the 2008 massive floods, never to be rebuilt.

There is no sign of people here now, however. The land is desolate and the buildings are burned to the ground.

Isaac asks, "Wonder if the Cushmans survived or where they are? Are they in hidin' also, or were they taken by the Gia Union Troopers . . . or maybe they succumbed to the firestorms and meteors got them? Their buildings are burned.

"You know, I met the Cushman family years ago when I was kayaking down this river. They were really attached to their land and had kept the estate intact. It was part of their lifeblood, part of their identity and character. I can't see them voluntarily leaving."

Ian says, "I didn't know them personally, but I once saw a documentary on well-known, local pioneers and the Cushmans were featured as some tough folks, for sure. They crossed the Atlantic in a Dutch cargo fluyt, settled in Massachusetts, then Vermont and then moved here. But hell, I think those things are nothing compared to the stuff that's been happenin' and the crap we've experienced the last three years."

Isaac says, "Yeah, you're right, but their ancestors did have their own rough time surviving."

Each man now has to push his pole into the soft pond meadow bottom as the raft is bottoming out in some places and the current is minimal. As he digs in for a big push, Aaron looks up to the right. The Cushman family cemetery is on the low-lying hill. One prominent marker is still clearly marked, but the letters are barely visible.

"Look," Aaron says. "Can anyone read that marker?"

Ian, a former Marine sharpshooter, has wisely kept his scope through all this. He whips out the scope and reads the grave marker.

"CYRUS CUSHMAN
HE WAS OF THE 8 GENERATION

FROM REV. ROBERT CUSHMAN
THE PILGRIM"

Isaac says, "It's funny . . . when I talked to them, Robert Cushman was best remembered by his family as the ancestor who missed sailing on the *Mayflower*."

Aaron feels the raft scraping the bottom and gives a mighty shove before responding. "That's weird. Half the country used to brag about how their ancestors came over on the Mayflower. Hell, a whole cottage industry was built around that. This is the first time I've heard of anyone boasting about missin' the boat."

Isaac replies, "Well, I got the whole story directly and, in many ways, I think their story has strong similarities and parallels many of the world's current situations.

Aaron says, "Well, sorry my friend, I think you're full of shit. Nothing in history has every compared to what we've been going through."

"Okay! I never said 'it compares to,' I said 'similarities.' Those people were forced to leave their homeland to avoid tyranny and religious discrimination. They survived hostile living conditions, relocations, natural disasters and spiritual upheavals before they had a chance to settle here and rebuild."

Ellis says, "Aaron, you should know you can't out-piss a skunk. When it comes to history, you just as well give in and listen to him."

Aaron shoves his pole so deep into the muck he actually slows the raft as he pulls it out.

"Alright, I guess I can endure one more of Isaac's long-winded lectures . . . since I don't have anything else to entertain me."

Isaac is famous for his allocutions and the other team members have learned to tolerate, and in fact, sometimes to actually listen, because they know he has such a command and respect for history.

Without further argument or hesitation, Isaac begins.

He may have missed the *Mayflower*, but Robert Cushman was really one of the founders of Plymouth Colony. It was 1617, and Cushman was a member of the Leiden community in Holland. The group had received permission from James I, King of England, to settle in lands contracted to the Virginia Company and Cushman was delegated to negotiate with the Company. He was successful but, before the Separatists could sail, the Virginia Company went bankrupt.

Now guys, Cushman wasn't a quitter. He already had the king's permission, but needed a new partner to replace the Virginia Company. So he scrounged up a friend, Thomas Weston, who organized some venture capitalists to underwrite the new project.

Those Separatists sailed on their historic journey late in the year of 1620, and who would believe it . . . they became known to us as Pilgrims. Cushman was committed to the venture, but the second ship, the *Speedwell*, in which he was to sail, started to leak so it had to return to port. There wasn't enough room for him on the *Mayflower*, so he missed history and the first sailing. But I'd guess that delay saved him from the first miserable winter when nearly half of the new colonists died of starvation and disease. Finally, in 1621, Cushman arrived in Plymouth, finding an impoverished and despondent community.

Robert Cushman was not only a great statesman, but a hell-of-an orator. He had used his skills for lively and compelling sermons. They were instrumental in inspiring his brethren to overcome the hardships, not give up, and to move ahead with their quest for a new life, in a new world.

The Plymouth colonists did so, and prospered. Cushman's family grew and prospered too. But, as with many of the settlers, after many years, his descendants eventually left Plymouth for new adventures. The Cushmans moved northward, to pioneer in Vermont. There, by hard work, cunning, and dedication, they continued to prosper for generations. But that prosperity came to an abrupt end in 1816. The climate of North America experienced a sudden and unexplained change which became intolerable and severe in the north. Thinking this to be a local phenomenon, many picked up stakes and moved west.

Unknown to them at that time, a year earlier in 1815, a whoppin' volcano eruption of Mount Tambora in Indonesia had blown one bastard of a crater. It was eleven miles in diameter and half a mile deep. That son-of-a-bitch ejected over four-hundred-million tons of dust and sulfuric gasses into the stratosphere. It was so much dust that the friggin' temperature of the whole world dropped and there was no summer season in America or Europe in 1816. That damn volcanic explosion was ten times stronger than Indonesia's other, more famous, Krakatoa eruption which occurred later in 1883.

So not only was 1816 without a summer, it marked a serious worsening of the "Little Ice Age," a worsening which was to last for three decades. Over those years, crops failed and life, especially in New England and New York, became increasingly difficult. Thousands took up roots and moved west, hoping to find some relief from the severe weather and famine. The Cushman family was among them.

There was a lotta' concern about the declining weather conditions. Then, there was a big meteor shower in New England. Damn! That did it! Triggered a big bible-thumpin' return to religion in what was then an increasingly secular nation.

The so-called "Second Great Awakening" swept the country, especially in New England, New York and the upper Midwest. Millions opened their Bibles and prayed, seeking answers for the terrible change in fortunes. Camp meetings by the thousands and calls for repentance became common. Millions pledged themselves to God, if He would spare them from the crushing famine and tribulation.

Three new and distinctly American religions became offshoots of that second Great Awakening. They were the Latter Day Saints (Mormons), the Millerite revival, which led to the Seventh Day Adventists, and the Holiness Movement, which later evolved into modern Pentecostal and Charismatic groups.

Back to Cyrus Cushman. He was young when he moved to Ohio. The weather wasn't much better, so in a few years he was off again to the southern Wisconsin territory. In 1842, he laid claim to almost 920 acres along this meadow at the bend of the Bark River. He busted his ass building a log home, then a dam, a water-powered sawmill and an estate. By the mid-1840s, the effects of the Tambora eruption had dissipated and climate was improving worldwide. But this was most appreciated by the westward marching settlers of North America. Winters were less severe. Summers were associated with improving harvests. More rain, less snow.

Just think, all that chain-reaction change was triggered by a volcano in Indonesia, half a world away! If it hadn't been--

Ellis considers the timeline of Isaac's story and interrupts. "It's interesting to notice that in 1844, while hundreds of thousands from the Millerite Movement were gathering together in prayer, anxiously awaiting the imminent return of the Messiah, right here, Cyrus Cushman is building this estate on the banks of this Bark River. That excites me. This is profound! Really profound. To me, at least."

The profound moment and the location sinks in deeply with the team, a groups who have just now witnessed the real return of the Messiah. "Profound," each one lets the moment sink in.

Ian breaks the silence and says, "Well this time, He really did come. The Millerites thought they had it all figured out, and thought they were ready. But here we are, we got totally surprised. That will require some reflecting."

Isaac figures he has made a convincing argument about the Cushman story having strong similarities and parallels with the world's current situations and continues to silently punt through the shallow waters.

Aaron understands and appreciates Isaac's position, but decides to remain quiet. They are good friends, but he is stubborn too, and not about to give in so easily.

The raft requires considerable effort to navigate a sharp turn right, over the narrow spillway and finally around the remaining concrete dam. And they are on the way again.

Ellis is first to sight the head of a painted turtle poking out of the water. "Unbelievable! Look, a turtle, a real live turtle!"

Then Ian spots another one on the bank. "Way to go! That's good, damn good! Even if it's only reptile life . . . it's still life and the first we have seen."

Isaac says, "Turtles are amphibians, not reptiles, asshole!"

"Nope, gourd-head, they're reptile. You may know your history shit, but you ain't no biologist!" Ian says loudly.

Aaron and Ellis grin at each other, they agree with Ian.

Isaac heavily shoves his punting pole with a little extra effort. "I'm still is not convinced, but I do agree, either reptile or amphibian . . . it's good news."

Ellis says, "Yeah, it's cool we're seeing some life, but reptiles are like cockroaches, very hearty, tough critters. Low metabolisms gives 'em an advantage when food's scarce and the environment's poisoned. Now, let's look for some real life . . . some frogs and salamanders, true amphibians.

"You know guys, frogs are very sensitive to pollution and environmental changes and they were on the decline for decades before the plagues and firestorms. They're even sensitive to light pollution, so with the sun and sky darkened so long, I question if any survived."

"We'll see," Isaac whispers.

Beyond the washed out dam, the river curves southward. Just beyond is where Duck Creek joins the Bark River from the north. The five-foot wide Duck Creek adds to the water volume, increasing the river's depth and flow. The river straightens after making a ninety-degree turn and they are now gliding southward at a very comfortable speed.

A few miles downstream the team comes to another hairpin turn in the river that marks the hamlet of Hebron. Hebron, in the past, was renowned for its dairy farms, dairy products and cheese production. The men can't help fantasizing about a cold glass of milk and a generous chunk of delicious Wisconsin cheese. Oh, their minds are flooding with old memories.

But there too, they find another one of the Bark River's many dams that were built and later demolished. The Hebron Bark River dam had also once driven a mill, but had ceased that operation prior to the demolition by the Wisconsin Department of Natural Resources. At the time of its removal, records showed that the dam was deeded to the prestigious Hebron Rod and Gun Club. But unfortunately, the DNR could no longer find anyone belonging to that club. So flattening and demolition went forward.

Isaac says, "You know, the DNR spent decades removing dams obstructing Wisconsin's streams to restore the rivers to their natural state. Man, was that avidly supported by environmentalists from the liberal University of Wisconsin over in Madison.

"At the time, I resented those lame, liberal, progressive, radical environmentalists. But right now, just think how we'd be doing if we had to pull this heavy son-of-a-bitch raft out of the river, around some damn dam and back into the river, again and again. So, looks like the freaking DNR unknowingly did us a big favor. Not that I'm giving them any credit thinking about us."

The team passes through the broken dam, through the hair-pin curve, and southward again. On the right of the river are the burned-out remains of the Bark River Historical Society Museum.

Isaac cannot resist. "Any of you ever visit that museum?"

"No, but I'll bet you did," Ian says. Without even a pause, Isaac is into one of his history lectures.

"It was a damn nice little museum for a town the size of Hebron."

"I'll take your word on that," says Ian. "What so special about Hebron?"

"It's here in Hebron where a guy named Harry Wintermute established his famous circus, the Wintermute Brothers' Show. That was way back in the 1880s. He traveled around the country, with a tight budget, exhibiting mainly family talent and kept his circus going into the early 20th century. He did have one bright star, a high-wire walker who performed with her family, she was called Bird Millman. Later she traveled with the Ringling Brothers Circus and eventually made it into the Circus Hall of Fame. You can see her history in Baraboo, or maybe it's in Sarasota. Well, that's if they still exist?

"So much has happened along this tiny, obscure river, which now appears to be our life-line to a new world."

Before Isaac can start another story, Ellis decides to change the subject. "History, history, hell of a lot of history! But shit, where are the other people? I'm feelin' very alone out here!"

This comment again leads to silence and contemplation. Now drifting effortlessly and quietly southward, the shadows of the evening are growing long. The air is becoming chilly as the sun's autumn rays slip further from the evening sky.

Isaac knows the river well and tries to visualize what lies downstream. "We're making good time now, but we should consider where we want to put in for the night. By memory, I think we should be coming upon a couple'a sandbars that might be good. Or, about a mile further along we'll come to the mouth of the Scuppernong River . . . and a bit further, maybe two-or-so more miles, we'll see the mouth of Whitewater Creek near Cold Spring. Which one do you guys want to pick?"

Ian asks, "Which one you think is best?"

"Well, Scuppernong will give us about an hour of daylight to set up and secure a camp. Whitewater, however, will put us a couple more miles downriver."

Aaron quickly questions the decision to put in on an exposed sandbar for the night. "What about keepin' low from Gia Troopers?"

Isaac says, "We've gone down the river for two days now. We've passed a friggin' Gia Union Command Center, gone past three towns, Sullivan, Rome and Hebron with no sign of human life, including Troopers. I'm growin' more confident that the Troopers are dead or busy elsewhere."

Ian says, "If any Troopers have survived the coming of Yeshua in Jerusalem, don't you think they would've already been summoned by Gog to fight back? I don't think that three days after Yeshua's coming, they'll still be here on the lookout for us. Well, least that's what I think. What about you guys?"

Ellis is first to reply. "I agree. We've fought long and hard, managing narrow escapes and experiencing miracles we didn't recognize until now. Someone up there's looking out for us. Still is. Aren't we on a mission now, a mission from God, with direct instructions from one of Yeshua's Kadoshim? I don't think he would send us into a trap. And, even though we haven't seen him in two days, who's to say that he's not been watching over us, even now."

Aaron voices a contrary view. "Yeah, I agree we've been lucky, maybe even spared by miracles, but we've used our training, our survival skills and cunning. Do you think that God wants us to let down our guard now? Seems I remember a verse from somewhere, 'He that endures to the end will be saved.' I think we need to keep our vigilance up."

The discussion grows more intense. Both sides are right. Both sides are wrong. The discussion is loud and intense. Obviously, there's a major disagreement, but the four have gotten along for all these years, together and as individuals. No rank or command structure has kept them in line or restricted them from dissent. Aaron, who has the highest Marine rank, no longer feels the need to press his echelon as a command structure.

The river is flowing faster, and the Scuppernong is visible on the horizon. The Scuppernong River is a real river, not one of the many creeks that contribute to the flow of the Bark River. In fact, the Scuppernong is the only real river tributary to the Bark. As their raft approaches the river's mouth and the sand bank at its entrance, there remains no agreement among the four men. They are increasingly venting steam, and yes, getting personal satisfaction from yelling at each other.

Then, with a brief pause in their loud arguments, a sound tweaks their keen ears. Voices? Human voices?

"Quiet! Shut up you jarheads! Quit it! Shut up! Listen. Whaddaya hear?" Aaron demands.

All four snap to silence. Then lean forward while shading their eyes and cupping their ears to pick up the distant sounds coming from the Scuppernong shore. Ian quickly retrieves his sniper scope and focuses on apparent movement on the bank, port side.

THERE ARE PEOPLE OVER THERE!

Ian says, "Hold on! There's three, four, maybe five on the bank. Looking this way. Yelling at us!"

"We're not alone!" Ellis shouts with a broad smile. "There're the others Yoni told us about! The others to join us? Maybe? Maybe?"

The raft is moving at a fast pace, three to four knots, and swinging outward from the sandbar into the more rapidly moving current opposite to the bar. As the raft approaches the bank, images of the second group become clearer. There are four women and one man on the bank, yelling at the raft, motioning for them to come ashore.

All four rafters push their poles starboard and into the river bottom, pushing the raft to the left, sideways out of the more swiftly moving water, and toward the slower flow along the sandbar on the port side. The current requires all the effort they can muster, but gradually the raft moves closer and finally comes to a stop on the sand bar.

The two groups look at each other with great anticipation and awareness. They are survivors . . . from the Gia Union, from the tribulations, from the trumpet-plagues, the bowl-plagues, the firestorms, from famine and from isolation.

Ellis' question, "Are we alone?" has been answered.

As the raft glides to a stop, the group on the bank moves slowly down the muddy bank toward the sandbar and raft. All the men on the raft move to the port side, toward those on the bank. Each group moves cautiously towards the other, without speaking a word.

Just before contact, about ten feet apart, one female voice brightly says, "Hello, or is it shalom? My name is Teah." She stretches out a welcoming hand.

Then there is a simultaneous rush of introductions.

"Ian"

"Rivka"

"Ellis"

"Michael"

"Isaac"

"Aaron"

 "Raquel"

"Miriam"

After these restrained, formal introductions, the two groups merge in heart-felt hugs and mingling. Each has many questions to ask, but all know that the answers will be the same. They are two groups, spared the destructions of recent times, and they are all on a common mission.

CHAPTER 6

HAR MEGGIDO

JERUSALEM — DAY 6

As the sun sets in the west, Yitzhak (Isaac) and Daniel, two Kadoshim working in Jerusalem, cross the Golden Bridge spanning the Kidron Valley and sit down on a rocky outcrop on the western slope of the Mount of Olives.

Daniel asks, "Enjoying the view? Or reflecting on the progress across the valley on the Temple Mount and Jerusalem?"

"A bit of both. And just thinking. It was nearly three thousand years ago when my father, Avraham, led me up that slope over there, right over there, and tied me down to that rocky outcrop. He had become convinced that he needed to sacrifice me there, for some reason unknown to him. And God did not intervene to stop him, until very the last moment. It's interesting to recognize how often God does that."

Daniel says, "I know your story well, have read it dozens of times. It was part of our Torah readings every three years."

"And your memories?" Yitzhak asks.

Daniel slightly changes his position so that he looking directly at the city walls as he replies. "I remember growing up in the royal household and the confusion and stress as the Babylonian armies took town after town, tightening the circle around Jerusalem.

"Then the long siege lasting a year and a half and the horrible conditions during that time. And the terrible slaughter when the walls were breached. Four of us hiding in a cistern as the city burned around us. And finally, being led away as a young man, chains on my legs, part of a long column, marching north through this very valley in 587 BCE. I and my three friends were captives of Nebuchadnezzar, King of Babylon. I never saw Jerusalem again, except in visions. I lived the rest of my life in Babylon. Oh, yes, I remember, in Babylon we sang a "non-song" about this place.

'By the waters of Babylon, there we sat down and wept, when we remembered Zion. On the willow trees there, we hung our harps. For our captors required of us songs, and our tormentors, mirth, saying, Sing us a song of Zion. But, how shall we sing Yehovah's songs in a foreign land? If I forget you, O' Jerusalem, let my right hand forget its skill. Let my tongue cleave to the roof of my mouth, if I do not remember you, if I do not set Jerusalem above my highest joy.' (Psalms 137)

"Indeed, decades later many did come back home and built another Temple which, in its turn, was destroyed by the Romans. Then there was that small recent Tabernacle, lasted less than four years, before being destroyed itself. Now, here we sit watching the foundation being laid for a third and final Temple that will never be destroyed."

The two continue to talk, listening and watching as the sun finally sets over the Jerusalem horizon.

Yitzhak says, "Listen. Listen to the songs coming from the Levites, before the altar. Listen to the echo through the valley. Restoring the Temple, the center of worship, is central to God's plan, putting Israel in its proper roll among the nations. From this site, God's righteous law will be taught and will go forth to all the world."

The whole mountain has required purification ceremonies. No red heifer was required to cleanse the grounds this time as the Messiah personally cleansed the priests, the utensils, the altars, lamps and the ark.

The urgent need to restore the population is well under way. Kadoshim have been locating people who fled from Jerusalem three years ago, assisting them in their return home, from their hide-outs in the mountains and caves of the Judean wilderness and even from the wadis and valleys beyond the Jordan. More are on their way.

Many are already eagerly participating in the reconstruction works. Marketplaces and businesses are beginning to reopen. Open-air shops, cafés, and small stores are appearing already.

And while the Temple work continues, homes within and outside the Old City are being cleaned and repairs have begun. Jerusalem is beginning to buzz again.

As the evening shadows fade and darkness shrouds the valley, the two Kadoshim continue to relax and enjoy the peace of Jerusalem.

Strange! Are they unaware? Are they oblivious to the massing of immense armies in Megiddo just sixty miles to the north?

HAR MEGIDDO – DAY 7

"Stan, gimme a hand with this laser-shield modulator. The Commander wants this place abso-fuckin-lutely secure. At all times. No lapses. No down time. No SNAFUs," Gary commands.

Stan gives his partner an exaggerated, sloppy salute. "Yes sirree! But this damn system needs a lot of tweaking to keep it functioning. Am I the only son-of-a-bitch who understands this technology?"

"No, but you are the top bastard, anywhere, to keep it operational," Gary says. "You know what? We are both tops. I don't know what the Commander would a done without us."

"Excuse me."

Gary and Stan are startled. They had not heard Warren Bates, the new senior technical advisor, come up behind them. They had met Bates only yesterday, along with a couple of other new arrivals especially deployed for this critical operation. Gary and Stan were not impressed with the civilian advisors, particularly Bates. They thought he looked too young to advise anybody, especially them.

"Excuse me, gentlemen," Bates says again. "Based on what I have heard and those self-evaluations, which I couldn't help but overhear, it sounds like you're sort-of-a-bit of hot-shit around here."

The newcomers' dismissive comment immediately puts both men on the defensive.

Gary's face flushes red. "What the fuck do you know man! The Supreme Commander was lucky to find us … a couple of combat-hardened, marine officers who just happen to have technical expertise in laser reflection technology."

"Yeah, Mr. Bates," says Stan. "The Gia Union recruiters wanted us really bad, and gave us a shit-load of promises to get us and we're worth it."

Bates says, "Whoa, man! Hold on. Didn't mean that. Guess I started off the wrong way. The Supreme Commander himself was just telling me how pleased he is to have you two as his personal guards and how highly he thinks of you. He suggested I should come down here and have you guys fill me in as to what has been going on. And just call me 'Bates,' no need for 'Mister.'"

Only a little embarrassment impinges on Gary's and Stan's super egos. They know they have it good and are not going to let any "civilian" bureaucrat get in the way. However, Gary feels he must make sure that Stan's comment on "promises" is not misinterpreted or incorrectly reported.

Gary says, "It's for damn sure we were promised a lot, but, I gotta admit, the Union Command has really honored its agreements."

Bates believes he is slowly overcoming the awkward situation. He was aware of their big egos, but had not anticipated their thin skins.

Bates says, "Fellows, we're very likely to be working together after we clear up this Messiah mess, so let's sit down, take a quick break and get to know each other a little better. Tell me how you ended up here."

Gary still feels a little threatened by the newcomer. He's been stung before by the duplicity of smooth-talking civilian consultants who would sell their grandmother if they saw any political advantage for their own career. However, he decides the best option is to not pick a fight and try to appease the guy.

Gary says, "Well, it just seems just such a short time ago when we were fighting against the Gia Union. We were in Chicago. But the time came to make a decision, and now here we are, on the other side. We've moved up quickly. Seems unbelievable, but we rapidly became two of the Supreme Commander's personal guards, responsible for his security. And that includes shielding him from advanced assassination weapons"

"Yeah, like, you know, those compact particle-beam guns them crazy Israelis tried," Stan says. "Yep, I think we proved ourselves then for sure. The Union needs our knowledge of laser-reflection technology, as well as for our military skills. We're good . . . and we've proven our loyalty."

Bates says, "I don't think anyone questions your loyalty. Your files show you were very eager to take The Marck."

Gary's anger and defensive attitude begins to subside. "You bet. For sure we were ready. What if we had not taken The Marck? What if we'd rebelled and had gone AWOL like those lousy deserters in our unit did?"

"Personally, I can't even imagine it," Stan says. "We've had it really good. No turning back. No second thoughts. Look at us. We've got power, honor, and respect. We live the life of a prince — well, warrior princes, for sure. Whatever we need or want's provided. Gog has been so good to us."

Gary nods his head in agreement. "Yeah he has, but he does depend on us for protection and we're good at that. So, I think we'd better get this screwed-up modulator back on line soon. Stan, you can work as we talk."

Bates stands up, brushes the wrinkles from his freshly ironed pants and looks at the two.

He asks, "Don't you sometimes wonder, what became of the rest of those traitors in your unit?"

Stan looks at Gary. "In the last report I heard, a couple years ago, they still had not been found and were still on the run. But who really knows. Without food, supplies, or communication there's little chance they've survived."

Gary says, "Well, you know, we were all then like brothers. Hell, why didn't they just join Gia with us? It's frustrating. The poor dumb-shits. But it was their stupid decision. We all gotta make our choices. And we made ours logically. We did the right thing. As I keep on a'saying, just look at us now, personal guards to the most powerful man in the world. We're highly respected everywhere. And we get to live in Jerusalem, the capital of the world. And I gotta say, we've had some unbelievable weekends at our personal condo, right on the beach in Caesarea."

Stan drops his guard and loosens up. "Hell, we're enjoying it all, including the wine, women and song. Gary, you can keep the song, just gimme th' women! Heh, Heh, Heh! Yep, soon as we get through this deployment, I'm heading – straight to the beach."

Gary thinks Stan is talking too much and steers the conversation in another direction. "I still don't understand why there's so many who don't appreciate what the Supreme Commander and the Gia Union have done for us all, for the whole world. Well, it's just too damn bad for them. They should've learnt a long time ago to make the right choices: to get in step with the program!"

Stan understands Gary's interruption. "Yeah, and remember them Jews. Talk about a bunch of ungrateful bastards. Remember the first Jewish riots in Jerusalem. I think that's where we really proved our loyalty to the Supreme Commander.

Gary says, "Yeah, I remember. When the protestors advanced onto the Temple Mount, we had to confront them directly and open fire on them. Hell, even then, they didn't stop. They just kept a-coming. We killed so many and still they just kept a-coming. The next day more were back, again. What were they thinking?"

Bates is restless. "Yes, that Israeli rebellion was so unnecessary."

"Yeah, why would they be so thankless to the Gia Union that had brought peace and prosperity to this troubled land?" Stan says. "Why were they so focused on who's in the Tabernacle's holy place? I just don't understand these ungrateful Jews."

Gary says, "That's when the Commander was so impressed with us that he assigned us to command the Gia Special Ops unit that fire-bombed them sons-a-bitches in the Tabernacle. We sure surprised them. And, Wow, did that whole thing burn bright."

"It was such a shame about all that gold in there, it just melted and flowed down into the pavement cracks. What a waste," Stan says.

"Trying to assassinate the Commander! Now, that ain't gonna happen, as long as he's got me and Stan protecting him. And he knows it," Gary says. "Even up here in the Megiddo Command Center, he still wants to keep us close by his side. Yep. It's a good feeling to be so needed. Even up here there's no telling of how many infiltrators there are in the millions that are assembling"

Bates tires of their boastful behavior and abruptly starts toward the door. He says, "I shouldn't hold you up. I've heard all I want to hear." He leaves.

Gary looks at Stan and shrugs his shoulders. "What's up with that guy? At first he acts like a prick, then he's all "Mister Nice," and outta nowhere he gets up and walks off."

Stan gives Gary a silly grin. "He said not to call him Mister."

"Alright, from now on we'll refer to the young Bates as Master!" Gary says. "Okay, times up, ya got that modulator up yet?"

"Yep, got her. She's now working like a charm."

"Good, that's done, check that off. Now let's get our update reports on the Troop build-ups and their assignments. The Supreme Commander will want his afternoon briefing soon."

HAR MEGIDDO – DAY 9

Viewed from the Gia Union Command Center, atop Har Megiddo, Gia Union forces from every continent are assembling in the broad and long valley below. They spread as far as the eye can see and pile up over the ridges of Nazareth to the north, Mount Tabor and Gilboa to the east and up onto the Mount Carmel ridges to the south. The sky is filled with aircraft of every kind. Colonel Gary White and Lt. Colonel Stanley Campbell are at Gog's side, providing for his security and offering tactical advice. He has confidence in their counsel. They have proven themselves in many campaigns.

The Supreme Commander straightens, pulls back his shoulders and turns to Gary. "Colonel White, give me a status report."

Gary gives a smart salute. "Yes sir! Today is day nine of 'Mission Retribution.' The buildup is right on schedule. Troopers, tanks, armored personnel carriers, 'copters, cruise missile batteries, attack and bomber aircraft, drones, artillery units and weapons of all kinds, including our tactical nukes are in place. All are now pointed south toward Jerusalem, the goal of 'Mission Retribution.' Troopers who could not get here over land are, as we speak, pouring onto coastal beaches via our Gia Cruisers, with beach-heads extending from Haifa to Gaza. More Troopers from other distant continents are arriving today via our Gia Air Transports. And thousands of 'copters are bringing even more."

Gog asks, "Have the forces promised by the north-eastern federations been delivered?"

"Yes sir," Gary replies. "They have committed their entire offensive ground and air arsenals with over a million Troopers.

They have also included their entire arsenal of laser and particle-beam artillery. It seems they now fully understand that "Mission Retribution" is for us all, for our system's future and our new civilization. They're on board."

Gog looks Gary and Stan in the eye, smiles and addresses the two in a less formal tone. "Again, I thank you two for convincing the north-eastern federations to make the right choice. You will be well rewarded."

"Yes, Sir!" They salute crisply.

The Supreme Commander asks, "Are all the needed Gia Union resources on hand?"

"Yes sir!"

Gary says, "This is the biggest assembly of military forces in the history of the world. And I must say, sir, also it's the best. Beyond what you can physically see out there, there are additional bombers, attack craft, cruise missiles and more 'copter assault units ready to join in from naval units in the Mediterranean and the Black and the Red Seas, as well as many based in surrounding countries.

"We are all go, sir, awaiting your signal."

"What's going on in Jerusalem? Is Yeshua preparing a defense?" Gog asks.

Colonel White responds for the whole council. "No sir, and that's real strange. There are no signs of any defensive maneuvers. No signs of military activity. They are going about, as we speak, preparing for Yom Kippur tomorrow, with no signs that they are even aware of our forces only sixty miles away. They are just interested in doing their stupid stuff.

"When you give the command sir, we will surprise the shit out of them. A total shock! Shock and awe! The mother of shock and awe. We will strike them fast and hard. They will not even know what hit them."

CHAPTER 7

ON THE ROCK

DAY 5 - WISCONSIN – CONTINUED

After the excited introductions, Isaac shrugs his shoulders and looks at his crew. "Well the answer's now obvious. We're to spend the night here at the mouth of the Scuppernong, restore our energy, review our supplies, join forces and get a good night's sleep."

There are many questions each member of the parties has for each other. Although Isaac's team is a cohesive unit, built on many years of service in the Marines together, it is apparent the Scuppernong group does not know each other as well. Teah seems to be a natural leader, but the group shows the lack of coordination found in Isaac's tightly organized band.

Isaac and his team quickly set up camp and roll out the reed mats to sleep. They have, in fact, five extra mats they made last evening. They set up guard watch posts on the perimeters, while the other team is passive and seemingly somewhat uncertain what to do next. Teah especially watches the efficiency of these new guys.

Isaac, finally aware of her interest in their activities, approaches Teah. "Who's your Kadosh?"

"Yoni, and yours?"

"Yoni." Both smile at each other.

"How about your group, do you have food, supplies and clothes?" Isaac asks.

"Yes, Yoni provided them for us two days ago. He told us to walk to this river junction and wait for others to join us. Looks like you're those others. Or are there more?"

Isaac says, "It appears that you all may be the first of several groups we may be meeting along the way. I think it's likely we'll bring some others on board.

"Some extra food might be a good idea, don't you think? We have some berries and roots we have gathered and bread, provided by Yoni. It should be plenty for all but, in the morning maybe we can search for more berries and roots to have some excess as we move along."

Teah says, "Well, our group seems to have plenty for a short term supply, but it certainly won't hurt to have some more. We still have thirty-two bottles of water we picked up at an abandoned Quick Stop. We've been using them very sparingly. They may be needed for some others too."

"Good," Isaac says. "Let's be sure to take them on board tomorrow. Now for tonight. We've been avoiding the Gia Union Troopers for so many years, we're still somewhat apprehensive, so we'll post two guards on rotating shifts. The others will sleep better that way.

"Do you think any of your group might want to volunteer for a shift?"

Teah is a bit surprised by the continuing fear lingering in Isaac's party, but she respects their conservative judgment and volunteers. "Just give me instructions and I'll take a shift. I'll also ask the others."

Teah gathers her group around her and discusses the night watch shift. Each is skeptical, but everyone still volunteers to take a shift.

Ellis has built a fire on the sandbar about twenty feet from the raft. It is a natural gathering place and provides some warmth against the chilly evening. The two groups gather around the fire to chat and to eat.

Miriam inquires first. "Tell us your story. Who're you guys? Where do you come from?"

As they eat berries, roots, bread and drink water, each team briefly shares stories. It is already late, they all are tired and it's time to set up watch rotations.

Aaron says, "As much as I want to continue this, I think we all need some sleep. We gotta full day ahead tomorrow."

"I agree," Isaac says. "Besides, sailin' should be smooth tomorrow, so there should be a good chance to get to know each other better."

Ellis adds wood to the fire. He and Rivka take the first watch. Then each two-person night watch team takes its turn. But a boring watch it is. There is nothing to report to the successor. All sleep well.

DAY 6

All awaken as the dawn lightens the sky. No Gia Troopers are to be seen!

They prepare for departure and are quickly off down the Bark River with all nine passengers, but there is still plenty of room. Yoni had correctly instructed the builders.

The river, from this point on, is wider and less winding and deeper. The water is now pristine and the sky is also clearer. The red-brown haze, which all have become accustomed to, is giving way to blue, with white fluffy cumulous clouds passing occasionally in front of the sun. Within an hour, they pass the mouth of Whitewater Creek and move on toward the junction of the Bark with the Rock River at Fort Atkinson.

"So," says Ellis. "Let me get this straight. You five are from all over the country, even around the world, with a common root at UW-Whitewater?"

"Yes," says Rivka. "We were students at UW-Whitewater when the trumpet-plagues, bowl-plagues and firestorms began. I was an exchange student from Brazil. Teah is from Ireland and Miriam from Australia. Michael is from North Carolina and Raquel is from Puerto Rico. We've all been at Whitewater for over six years. Each of us came as undergraduates and continued for our graduate degrees there. We loved UW-Whitewater."

Ellis asks, "How come you guys got together?"

"Well, we all had shared classes and had known each other on campus, but one thing really brought us together – well the biggest thing. None of us had accepted The Marck when it was introduced to our fellow students.

"The liberal, tolerant settings of the University system allowed us to be endured, and in fact, we were even helped by some others who had The Marck. Those of us who didn't take The Marck kept as low profile as possible. But eventually we discovered each other and began to meet periodically in secret, to discuss options and stuff.

Ellis says, "I understand. Rejecting The Marck was risky. That's when our lives really turned upside down."

Rivka nods, "With time and growing acceptance of the Gia Union government among college liberals and progressives, things began to get tougher. After all, there was worldwide prosperity and big government was part of the progressive agenda. And besides, the Union was giving lavish educational and research grants to universities worldwide. Tuition was free. The professors all had funded research projects, and funded, tenured professorships were becoming the norm for most full-time professors. Student activities, arts and sports facilities were constantly being upgraded. Life, at least for the first three years of the Gia Union, was quite plush.

Michael has been listening to the conversation and joins in. "Lots of students volunteered to join the Gia Union Trooper Cadet Program. College expenses were already essentially free, but being a student GU Cadet gave them a good income, above tuition and expenses. Cadets were only obligated with one day a week active, rotating service, and later with three months' service in other parts of the world. Golly, that time was credited toward a second major in Gia Services.

"As a college student, although officially just a Cadet, the entry rank was Second Lieutenant and you were immediately commissioned as an officer. Service of only fifteen years could to lead to retirement at seventy percent pay and a guarantee of a second career in another Gia Union government service. As you can figure, colleges and universities like ours served as a fertile recruiting ground for Gia Union Troopers through the Cadet Program.

"Gee, even faculty and university administrators joined the Gia Forces Reserves, serving one weekend a month and two weeks per year."

Rivka says, "So, the Gia Union Troopers are not just a bunch of common thugs. Despite the atrocities they have committed, they began as normal people, often intelligent and highly educated. But, something did seem to change when they donned those uniforms."

Ellis shakes his head. "Well Rivka, I'm not sure I completely agree with you, but it does sounds like the Gia Union didn't need to play hard-ball in the universities."

A small breeze has picked up, making the river trip more pleasant. Rivka brushes a pesky strand of hair from her face before replying to Ellis. "No, they didn't. It just seemed obvious to all liberal and progressive students, faculty and administration that the Gia Union was good . . . for the university, the people, the nation, the individual and for the world. Of course, speech against it was ridiculed and discouraged by administration, faculty, students and everyone it seemed.

Michael asks, "Why wouldn't anyone want to join the Gia Forces, at least as a reservist? On the surface it looked so logical. And why anyone one would refuse to have The Marck inserted was considered just plain stupid or antisocial. Maybe even a sociopathic, like one of them preppers we learned about. You know, many years ago, those weirdos who refused to have a driver's license, a social security card, a credit card, or a passport."

"Yeah," Ellis says, "I've heard about those guys."

Rivka looks at Michael. "Everyone knows, college campuses have always been a place for rebels to hide out and be tolerated, but at a Wisconsin state university, tolerance and diversity is a hundred-and-sixty-year-old tradition. That tolerance and acceptance, however, began to change after the trumpet-plagues started. The Union realized there was a serious adversary out there. There was an enemy seeking to overthrow the so-called beautiful world system it had so brilliantly established.

Michael says, "That's when the Union began to take a more active role in college affairs, actively seeking to identify dissenters, especially those without The Marck. I've heard it was worse at the big and famous schools and major state universities, the Ivy League schools and places like Duke, Vanderbilt, Emory and such. But fortunately, the more under-the-radar, rural schools like UW-Whitewater were further down on the hit list. Yet, we were still large enough that we, as individuals, could fade into the woodwork somewhat, with minimal attention."

Rivka smiles as she remembers. "Other students provided us with food and clothing and needed goods. There were a growing number of empty dorm rooms from those that left school with new Gia Union Trooper deployment orders. So, we were able to move from room to room when people or authorities began to close in."

Ellis asks, "How long did that go on? Did you ever get caught?"

"Well, it stayed that way, a game of cat and mouse, until the last fireball disrupted all campus electronic communications. All electronics, computers and records were lost. To our delightful shock, we gladly learned that the GPS system in The Marck software was severed too. Wow, the Union could no longer track its population. Furthermore, all Marck-monitored transactions were lost. And as for UW-Whitewater, the school really didn't have any option but to just unofficially shut down."

By this time, Teah has moved closer to the small group and joins the discussion. "Before the school shut down we had gotten to know each other well and kept an eye out for each other, but we had to keep separated to minimize suspicion. Some time ago when we were having a rare breakfast together, reporting and updating each other, two Gia Troopers walked right up to our table. 'Hey, you all do not register on the computer list. Let's see your Marck,' they demanded.

"We were stunned! What to do? Then Rivka recognized one of them from the UW-Whitewater co-ed lacrosse team. She stood up to him –."

Rivka says, "I looked him in the eye, smiled and said, Okay, Tim, and let's see yours!"

Teah laughs. "The Troopers were taken aback and looked at each other. Then, Tim smiled back and replied, 'See you later, Rivka.' The other Trooper smiled at Miriam, who smiled back at him and gave him a wink. He had a big grin as they walked away. There were numerous such close calls, but somehow we got through all that on campus."

Isaac now joins in and questions Teah. "So, Teah, why did you all not take The Marck?"

She pauses and each looks at one another for several seconds.

Then Michael, speaking for the group, answers Isaac. "We all kept asking ourselves that question too. Why not? It seemed the logical and maybe even the right thing to do. But within each of us we had a voice saying, No! Even though none of us have recently been active in a formal religion, I think that a little voice from our past training would not allow us to do what looked to be so practical, so logical, and what would have been compliant.

"All of us had some religious training in our childhood, and we could vaguely remember hearing of a coming evil beast power with a mark, sealing the future of all who take it. But we did not remember much else. We just had a bad feeling about it. So, Teah decided to research it some more and she went to the library and stole a Bible, didn't you Teah?"

Without hesitation, Teah speaks up so that all on the raft can hear. "Yes, I thought we needed more information and remembered a presentation I'd heard in an undergraduate course I had taken, Religions of the World. Everything that was going on brought to mind the class discussions about certain sections of the Bible. I felt that we just had to get that book for some serious study. I did not want to check it out because that might have raised suspicions. So, I secretly sneaked it out of the library.

"Later, I was really stunned when the Bible fell open at Revelation 13, telling about a mark that an antichrist will require of people. It described a mark, without which people would not be able to buy or sell. And it was to be in one's hand or forehead. I immediately showed it to the others. That was the clincher."

Rivka says, "There were no doubts after that. A prophecy from two thousand years ago was speaking so clearly to us today. So, when the school shut down we put our heads together and developed a plan to get away. We gathered as much loot as we could find and carry. Together, as a team, we moved off campus, slowly making our way over to the banks of the Scuppernong River where we followed its course north toward the Bark."

Rivka continues to fuss with her hair while she talks. "We started reading the stolen Bible, although we have not yet finished it."

Teah says, "We are sort of reading it backwards, you know, the last part first. Maybe that's our Hebrew heritage. Did you know, Hebrew goes right to left?" She smiles and winks as she is trying to entertain the Marines.

Isaac says, "Well, some of us do."

Rivka seems satisfied with her hair and goes on with the story. "Well we continued to search for food from abandoned stores, gas stations, and vending machines. We found three prized cases of bottled water along the way. We were getting low and running out of supplies, as there was less and less to find. We were getting weak and dehydrated, but we knew water was a critical element and we've been using our store very sparingly. Our lungs were full of the toxic air and we all have had chronic hacking coughs."

Teah says, "Then we saw the hologram-scroll in the sky showing the return of Yeshua. We heard the shofar sounds I remembered from childhood. You all must have seen and heard that too? Then Yoni showed up. And here we are today."

Aaron asks, "Were there other students at Whitewater who did not take The Marck?"

Michael says, "Yes, but it was kept as sort of a private matter, not something that you bragged about. There was no communication among those at other universities, and no formal communication at Whitewater. So, we don't really know how many, but think it was probably several hundred at first."

Rivka says, "We personally know of over a dozen more students that we had individual contact with, but they simply one by one disappeared or faded away. We lost contact. Only we five are left, to our knowledge. We hope that some of the others may have escaped too. But the prospect is not good."

There is a pause in the conversation, everyone is looking at each other, trying to quietly digest all of this.

As the raft now begins a series of hairpin turns to the right, then to the left, attention soon shifts to maneuvering the bow and stern around the bends. The new Whitewater team takes turns with members of Isaac's team.

Ian says, "You all catch on fast, for college students."

The Whitewater students pretend to ignore him. Each man rotates with a member of the other team, giving relief to all. Around another bend and the Fort Atkinson suburbs are now on the horizon.

Aaron says, "This will be the largest city we've passed through to date. It has major roads transecting it and there will likely be a Gia Union Command Center there. Let's keep as quiet as possible."

Tension builds. Everyone is quiet. Conversation and commands are at a whisper level. They pass under a bridge and enter the city. Burned-out buildings and vehicles are on both sides.

Isaac softly says, "Sure enough, there it is."

A conspicuous Gia Union Command Center looms ahead on the starboard side.

"But I don't see any activity around it. Nothing is moving."

Still everyone remains quiet until the raft glides safely past the Command Center. Then they all turn toward Isaac as he makes an unexpected, loud announcement. "Ladies and gentlemen, it's now time to say goodbye to the Bark, we're about to join the much bigger Rock River which will eventually take us to the great Mississippi."

Then suddenly, loud voices!

"Hey, guys!" A voice rings out.

They have been spotted! Everyone drops down on the raft. All keep quiet and a low profile on the deck.

Teah is the first to notice another raft tied up on the bank of the Rock River. About a dozen people are on or beside it and a second raft is tied up behind the first. And over on the bank, it is none other than Yoni!

"Hey, guys! What took you so long? We have been waiting for you all day!" A voice from the group inquires.

Everyone comes to their feet, and taking deep breaths call out in unison, "Hello! Shalom!"

Isaac and his nine-number team pull along-side the first raft and tie to it. Everyone, in single file, walks across the other raft and steps onto the shore of the Rock River.

"Good-bye Bark! You've been an adventure!" Ellis declares.

Yoni introduces each member from the three rafts to each other, while giving a succinct two or three-line introduction of each person.

Yoni has again prepared fresh bread and looks at the group with a twinkle in his eye.

"And I have another surprise," he states. "Wine." He holds up four large glass flagons of wine. "Later this evening we will make camp and enjoy a Sabbath rest. Some fresh challah bread, and wine will give us a great start.

"But for now, the day is only half spent. Let's see how much water you can cross before evening. Away!" Yoni is off as quickly as before, fading over the southern and western horizon.

Isaac walks over to Joel and Matt, the designated leaders of the two other rafts. "Should we tie the rafts together for the trip, or one follow the other?"

The consensus is to keep the rafts separate, as the Rock River has its narrows, rapids and bends. After a quick snack of berries, roots and fresh bread (the wine has to wait), and a little time for personal needs, as required, each raft pushes off into the Rock River stream. Fort Atkinson appears to be uninhabited as the rafts drift by the southern suburbs of the city. They pass the historic site of old Fort Koshkonong on the port side.

Isaac cannot pass by these historic sites without his compelling history lesson. "On the port side over there is the historic site of Fort Koshkonong. It was built by General Atkinson's troops during the Black Hawk war in the 1830s. The city used to hold an "Old World Remake" for us history buffs. I once spent a Sunday there, talking about life in the 1840s. We cut some logs with a two-man cross bow saw and there was a log splitting contest—"

Ian asks, "Did you win?"

"Hell no! Not even close. I'm a better historian than lumberjack."

"Roger that," says Ellis.

Isaac decides to ignore Ellis this time. "The employees and volunteers were dressed in nineteenth century frontier costumes. They even had a man dressed like Black Hawk himself, although, I don't think Black Hawk would have been appreciated there, back then."

Teah says, "I'm from Ireland so I don't know much about that war. What was that all about?"

"That event, the Black Hawk War, has remained a continuing controversy for historians and anthropologists. The ethics, the policies, the intrigue and the characters have kept this little known part of American history alive. The characters included such notorieties as General Atkinson, General Winfield Scott, Colonel Zachary Taylor, Lieutenant Jefferson Davis and Private Abraham Lincoln, to name a few.

Teah asks, "Abraham Lincoln? I thought he was part of your Civil War."

"Yeah, you're right. But before that he with served under General Atkinson"

"Was he here at this fort?

"Sure was! Now the Fort was built at General Atkinson's command. Later Congress honored him by naming the city after him. The General was the man who took the credit for defeating Black Hawk. Private Lincoln did not actually participate in the building of the Fort, as some have claimed, but he did serve as a scout for General Atkinson. He scouted along the south bank of the Bark River, searching for Black Hawk as far as Whitewater Creek. But, Black Hawk was on the north bank. Lincoln did not find him, for which, Lincoln later said, he was thankful."

Teah is listening intently, amused and fascinated by Isaac's command of history and his knowledge of the local area. Isaac, noticing a new audience, leans back and continues with details of the Black Hawk war and its effect on the history of the United States.

"Black Hawk had a brother, Neapope, a medicine man, who was known as 'the Prophet.' Neapope's prophecies provided the directions that Black Hawk needed in deciding matters of war and treaties. After the war, Black Hawk and his brother 'the Prophet' were taken on tours around the country. Paintings of them were commissioned, for payment of course. Even President Jackson, himself a famous Indian fighter, received them as honored guests and obtained a personal portrait."

The other seven are listening, but with less intensity than Teah, to whom Isaac is unconsciously directing his verbosity. All the others are remaining quiet, gradually redirecting their thoughts about what is happening, and how rapidly things are developing.

Teah, finally also having enough, interrupts Isaac. "Say, I understand that you have a Siddur. I haven't seen one in ten years. Would you mind if I took a look at it?"

"Sure, looks like there is some growing interest in these prayers. I've have been going over them too, so maybe we can study some together."

"Yes, I'd like that." Somehow, Teah feels both comforted and comforting.

The cliffs along on either side of the river gradually grow higher, the limestone hills having been eroded by the bigger river.

Ellis is again first to notice promising signs. "Look, see those gourd-like hangings. They're nests of cliff swallows."

Within a few minutes all can hear chirps coming from live birds in and around the nests. As the rafts pass by, several swallows dip down toward the rafts, swooping down and upward, again and again.

"More animal life!" Ellis says loudly for all to hear. "And birds are sensitive to the environment!"

There is a big "Hurrah" from all three boats and then an "Ooo rah!" from the Marines.

Just ahead is Lake Koshkonong. The plan is to spend the evening and the Sabbath tomorrow, resting on its shores. As the rafts move into the lake at the Rock River's northeastern entrance, the river's mouth and the adjacent land is marshy, filled with dead reeds mingling with budding green grass shoots. Although not a good place to disembark, the presence of plants growing again is encouraging. This is another good sign.

Ahead on the left is a sandy beach. That is the obvious place to land. All three rafts pull up on shore. Ian and Matt lead the search teams for firewood and more edible berries and roots.

Ellis becomes excited as he sees several fish jumping out of the water, splashing above the surface of the lake. "We're going to have fresh fish tonight too … if we can figure out how to catch one!" he exclaims.

Rivka says, "Wait! I've got a couple of fish-hooks and a line in my kit! But what about bait?" Everyone is stumped.

Aaron and Miriam light the fire on the sandy beach and soon Ian and Matt are back with broken tree limbs from the near-by woods.

Isaac begins a lecture about Lake Koshkonong.

You won't believe the environmental mess the US Fish Commission made when they screwed up and introduced carp, an alien species of fish, into the lake. I just can't imagine why the government would do that on purpose.

You know common carp, Cyprinus carpio, are native to the Caspian Sea. From there they migrated in to the Aral, then the Black Sea. Carp eventually spread east into Asia and to the west up into the Danube River. For some reason, carp were and are still considered a valuable food in Asia and Europe.

In the 1930s they were imported by Americans for private ponds but some were even released into the nation's waterways, including Lake Koshkonong. German immigrants to America were delighted by the availability of carp. They were eager for a steady supply. Thousands of carp fry were often dumped into rivers from railroad trains and cars running adjacent to the streams. Carp flourished, especially here in Lake Koshkonong and the adjacent Rock River. Before that, this lake had been a major commercial fishery for half a century.

But eventually, the carp displaced the native fish like our perch, bass, sturgeon and catfish. Carp are bottom feeders, stirring up the lake bottom, muddying the water, and most importantly disturbing the egg beds of other fish attached to the bottom growing plants. Soon it was almost all carp.

Finally, the government and fishermen realized the negative impact of carp and tried several unsuccessful strategies to rid the lakes of them. One scheme was to pull large seine nets between two boats. Large catches were obtained, but the carp still continued to dominate the lake.

However, the effort was beneficial for several businesses. I read somewhere that one entrepreneur promised to deliver twenty tons of carp on ice to New York City daily. As you know, carp is a favorite for gefilte fish on Passover and holidays.

It's strange, but some kinds of carp are famous for jumping into the air when the water is disturbed, you know like from a motor boat or oars. Carp are tough fish, but have they survived the toxic atmosphere, acid rain, meteors, fires and plagues?

Teah, realizes it is time to cut Isaac off. "So, let's go see what we can catch."

Raquel asks, "If they're famous for jumping, maybe we should use a net and some splashing, rather than a hook?"

Three of the women link their scarves together and wade into the lake, repeatedly dipping them as a net. And voila, up comes a six-inch silver carp! The successful effort leads to everyone joining in the fun, or cheering from the shoreline. Then the men take off their shirts to try the same process, but not as successfully as the women. Someone notes that when they splash the water surface, the silver carp will jump out of the water. In addition to some serious netting by the women, soon a game occurs of hitting the water surface and then trying to catch a fish in the air ... an effort with more fun than success. Actually, Matt and Ian do catch one each.

Within half an hour, over three-dozen nice size carp and two other fish, a bass and a perch, are hauled in. Now for the cleaning and preparing. The Marines, having multipurpose knives, are up to the job.

It's time to wash up. Ladies are first, and then it's the men's turn. No soap and no towels, however, just air-dry.

By the time all this is done, the sun is beginning to dip into the western horizon. Everyone stands and watches.

"It seems like a long time since we have seen a clear, golden sunset through clean atmosphere," Raquel says. And it has been.

Ellis has again found three big pieces of driftwood for makeshift tables. Each one reclines around them as the fish are roasted over the open fire.

Suddenly, in the southern sky, they see a light coming their way. It is Yoni coming to join them for the Sabbath meal, as promised.

"*Shalom aleichem*! Peace to you all," he declares.

"Now let's see. You have a fire, good. You have some wine and challah bread, good. And you have gefilte fish, well, okay. But you have no glasses, no candles or vegetables."

Then with a smile, he opens the bag he often carries and lifts out two candles with holders, thirty-four wooden wine goblets and fresh tomatoes, onions, carrots, potatoes and some brie cheese.

"I had to go to considerable trouble to get these, but you are worth it, I think." He grins and gives his now familiar wink.

"Let's get started."

Quietly, Aaron turns to Ian and tentatively asks, "Ian, are you feeling a little out of place here too? So many of the group seem to have Jewish backgrounds, and that's okay with me, but I'm still not sure why we're included in the group."

In almost a whisper Ian softly replies. "Yeah, I keep wondering the same thing. You know, early on I learned how to be an altar boy, and I know how to take communion, but

I've never really had a Sabbath eve meal and celebration. Sure hope I don't screw up and do anything embarrassing."

"Oh, I'm bettin' that you will," Aaron says.

Although Yoni is not close enough to hear them, he turns, approaches them and speaks just as if he had been a part of their conversation. "Fellows, you mustn't feel awkward. Remember, you were rescued and healed and included in this mission just as others have been.

"Already our small group of thirty-three represents a wide variety of nationalities, backgrounds, and upbringings. As we travel and others join us, you will see, our diversity will continue to expand and include cultures and ideologies, some which could be new even to you well-traveled men. From these, many beautiful and meaningful traditions will add to this New World. And be assured that none have been chosen for this mission randomly. All have an important place. Everyone is important.

"Just don't screw anything up, Ian." Yoni said with his characteristic smile and a wink.

Yoni turns to face the entire group. "We are all here together on a common mission with a unified goal, and first I invite you to relax and join me in an age-old, traditional practice of expressing gratitude and honoring Yehovah."

Aaron and Ian, feeling more relief than embarrassment, thanked Yoni and rejoined their companions with increased certainty. The group would be having their first Sabbath eve dinner together. It would be the first of a long series.

Rivka knows the blessing for lighting the candles and volunteers to lead the candle lighting. Isaac offers her his Siddur, but she declines.

"I got it," she says.

All the women stand and gather around the candles, set on a driftwood table, as Rivka leads the traditional Sabbath opening prayer for lighting the candles.

"Baruch Atah Yehovah, Eloheinu melek ha'olam. Asher kidshanu b'mitz vo-tav v'tzivanu.
Li-had-leek ner shel Shabbat kodesh."

"Blessed are you, Yehovah our God, ruler of the universe, who has sanctified us with his commandments and commanded us to kindle the light of Sabbath."

All say, "Amen."

Yoni pours wine into each cup. Each holds his cup high in his or her right hand, as they all together sing,

"Baruch Atah Yehovah, Eloheinu, melech ha'olam. Borei p'ri hagafen."

"Blessed are you Yehovah, our God, ruler of the universe, bringing forth fruit from the vine."

Yoni then breaks the pieces from the challah bread loaf, giving one to each. Each holds it high in their right hand as they sing,

"Baruch Atah Yehovah, Eloheinu, melech ha'olam.
Ha'motzee lechem min ha'aretz"

"Blessed are you Yehovah, our God, ruler of the universe, who brings forth bread from the earth."

All say, "Amen."

"Men, women, let's dig in. Everyone has earned a good meal and should have hearty appetites. Bon' appetite!"

As they eat and talk, even though having met only yesterday or today, they are already feeling like family. And Yoni is like a father and a mother and an older brother, all wrapped into one.

After each has eaten to the full and leans back to breathe deeply, Yoni reminds them of a special principle of thanksgiving found in the Bible.

"Isaac loan me your old Bible," he asks. Isaac whips out the old Bible and hands it over to Yoni. Yoni continues, quoting from Deuteronomy 8,

For Yehovah your God brings you into a good land, a land of brooks of water, of fountains and depths that spring out of valleys and hills. A land of wheat, and barley, and vines, and fig trees, and pomegranates. A land of olive oil and honey. A land where you shall eat bread without scarceness. And you shall not lack any thing in it. A land whose stones are iron, and out of whose hills you may dig brass.

Now when you have eaten and are full, THEN shall you bless Yehovah your God, for the good land which he has given to you. And beware that you do not forget Yehovah your God, by not keeping his commandments and his judgments and his statutes, which I have commanded you today....

But you shall remember Yehovah your God, for it is he that gives you power to get wealth, that he may establish his covenant which he swore unto your fathers, as it is this day.

After this, Yoni leads a blessing to God for the meal, for his protection, for all his good gifts, for his plan of salvation, and for his kingdom, for the Kingdom of God, which has now really come to earth.

And they all say, "Amen."

Yoni goes on to explain that this and other passages from the Torah are reminders of the source of our power, our wealth, and even our very food. And to think of them especially when we are satiated. And it has become a custom to do a thanksgiving blessing, after a meal especially.

"Our fathers seemed to forget God when they were doing well, only calling back to him and repenting when things are going badly. But this is not unique to Israelites, it is human nature. I think we can say the same thing about the Americans, Brits, Australians, Brazilians, and especially the Irish." Yoni looks at Teah and winks.

She smiles, understanding the gesture.

Then he says, "But we can do better. And we will do better. We are about to enter the most bountiful time the world has ever seen. So I think it's a good time to start restoring this concept in Israel, in fact, in the whole world. That's my Sabbath message."

There is a moment of silence.

"But," he asks. "Are you interested in a world news report?"

"Ken," "Yes," are the replies from the gathering.

Yoni continues with a progress report from Jerusalem. Then he gives a report from around the world. They learn that, at this very time, thousands of the remnant of Israel, like those here, are assembling for a return to the land of their fathers. From Asia, Europe, Africa, Australia, India, North and South America, the isles of the Caribbean, the Pacific, from around the world, from the four winds they will gather. Today, around the world they all have stopped, like these thirty-three travelers, to take a Sabbath break and to be refreshed for the long journey ahead.

Yoni reports, "Gog, the Gia Union Supreme Commander, left the city of Jerusalem unprotected, going north initially to attack rebels in the north and east. But now he is redirecting and assembling all his armies from around the world, assembling near Har Megiddo for a counter-attack on Jerusalem."

Yoni tells of Yeshua's progress in Jerusalem, cleansing the Temple Mount, starting the new temple project, and restarting businesses in and around the city.

"But," Yoni says, "The big battle, the mother of all battles, the Battle of Armageddon, of Gog, of the valley of Jehoshaphat, is pending just ahead."

"We need not worry, however," he concludes. "It will be taken care of. You will see."

He reassures them that all the Gia Union Troopers who have survived are assembling with the Supreme Commander in the Jezreel Valley, or are on their way there. He interrupts his report to state, "So, Marines, you can now discontinue your night watches, unless you just like staying up at night," he says with his characteristic grin and wink.

Everyone smiles, with a few short chuckles. Teah smiles at Isaac.

He adds this admonition. "Your immediate and compelling job is to get to the land of Israel, where much work will be ahead. There will be rebuilding, resettling and restoring the land. Yeshua wants you to be a part of the restoration yourself. He could do it all Himself, but that is not the kind of kingdom he has planned. You are to have a part. You are to be a part. Yeshua practices a way of giving. He gives to all the blessings of this wonderful world. And he gives and shares the joy of making it come to pass, the joy of work, of success, of the fruit of your labor. These are fruits which you can then give to others who will come afterward.

"Now winter is coming and there is still a long trip ahead. We will soon be travelling down the Mississippi, through the Gulf of Mexico, across the Atlantic, through the Mediterranean and to the shores of Israel."

Ellis says, "I think we are going to need a bigger raft!"

Everyone enjoys the humor and there is a sustained resounding laughter and chatter.

Then Yoni says, "I think so too!" And all the laughter resumes.

Yoni excuses himself and quickly departs off into and beyond the southwestern horizon.

The teams talk and question and listen to each other long into the night, around the driftwood tables and the fire on the beach.

Eventually, as all are retiring to their makeshift bed-mats, Teah asks Isaac if he would like to take a walk along the shore. Isaac is, of course, pleased to do so. They wander along the shoreline for a half hour or so.

"When we get to Israel, I hope we are not going to be segregated or separated. I don't know anyone in Israel," Teah says.

Isaac, unlike his usual personality, understands, but is speechless. He cannot get a word out. His throat is in spasm. He feels unsure what to say. Finally, after a long pause and some deep breathing, he blurts out a response. "Me too."

WISCONSIN – DAY 7

Morning comes all too soon, again. Several of the members are saying morning prayers and several are walking along the shoreline. Some others are sleeping in.

Then quickly there is Yoni again. He summons all to a breakfast of fresh fish (not carp), onions, tomatoes, olives, cucumbers and cheese.

"Where in the world did you get this?" Matt asks.

"Yes, that's right, somewhere in the world," Yoni smiles.

It is a pleasant, relaxed and unstructured day. A day of rest. There is no longer fear of the Gia Union Troopers, at least not here. There are growing expectations that this mission will be a success. How this is going to work out is becoming a bit clearer, somewhat.

Some gather to talk or to ask Yoni more questions. Some just listen-in to the others' conversations. Some stroll along the shoreline. Some take an afternoon nap. Eating, snacking, resting, talking, or enjoying another cup of wine. And after the afternoon meal, most slip into bed early.

WISCONSIN – DAY 8

Morning dawns. Yoni is gone.

"Well, we know what to do," Ian says.

After a quick snack, the leftover foods are packed, as well as all the supplies. Then the rafts shove off the shore into Lake Koshkonong.

It is several miles to the southern end where the Rock River flows out of the lake. Koshkonong is very shallow, four to six feet on average, with a soft murky bottom. They find there is no real current through the lake. And it is now painfully apparent, a fact no one had thought of before, they no longer had the flowing stream they previously depended on to power the rafts.

Crossing the lake requires constant pushing off the bottom with the long poles which have to be pushed hard, making them sink into the muddy bottom, getting stuck. The novice rafters realize they are not making much progress because pulling the pole out in the same direction pulls the raft backward. After experimenting, they soon discover a system of pushing in from mid-ship, with the pole angled to the stern, then walking forward, and pushing down and backward with one's feet as one pulls the pole out of the mud at a vertical position. It is taking quite a while to get this rhythm going. But there is plenty of practice ahead.

Despite the increased effort to punt the rafts through the motionless lake, there are enough crew on each raft now to frequently rotate the punting, keeping each person fresh and keeping the convoy moving forward. It takes most of the day to get to the Rock River outlet to the south.

Isaac is relieved to be back on the Rock. He yells, "Man, it's good to be moving with the river's current again."

And just when everyone has sat down for a breather, someone on the lead raft shouts, "Oh' No! Just ahead! A dam!"

Not Isaac, but Joel, is the expert on this. "That's the Indianford Dam, built a long time ago to raise the water in the lake."

The three rafts move along the eastern shoreline of the Rock River, exploring for the spillway that all hoped would be passable, or at least to find the best way to get by.

Joel says, "There's no freaking way to float over the dam or the spillway. The only way will be to drag these damn rafts out of the river up there on the eastern end of the dam. Then we can then work them around to the other side and re-launch below the dam."

So that is the plan. All become a little apprehensive since the rafts are very heavy. Too heavy for one crew to lift and drag.

However, Rivka quickly says, "Thank goodness we have three teams."

Here is where group cooperation provides a solution; enough manpower and womanpower. Everyone does his or her share of pushing and pulling Matt's raft first, until it was up out of the water and onto the sloping bank. That was the hard part. Then it was dragged up the incline, over the peak and then downward to the shore of the river below the dam. It was allowed to float downstream for about thirty yards and tied down there, to be clear of the others. The procedure was repeated for Joel's raft, and finally the heaviest of all, Isaac's raft.

It is a considerable effort to get Isaac's raft up out of the water and onto the steep shore. Then ahead, there is the steep embankment.

Matt says. "Dammit Isaac, your raft's nearly twice as heavy as the others because of these large logs stabilizing each side."

Isaac attempts to be positive and lighthearted. "Yeah, but it rides smoother too."

Finally, Miriam yells, "Why don't we reduce the friction by pouring some water on the shoreline and embankment?"

"Why didn't I think of that," three men say simultaneously.

And it works. The raft slides up onto shore and up and across the wet embankment. And of course, sliding downhill is easy.

Joel says, "We're on our way again … but there is only an hour remaining before dark. We need to remember this is called the Rock River. With the rapid current, sand bars and rocks in the river, it is important to see ahead. No way can we raft in the dark here.

"Up ahead, on the outskirts of Janesville, is Riverside Park. It used to be real nice. Hell, it used to have grassy shores, a boat ramp, baseball diamond and golf course. Let's push hard and shoot for that."

Ellis says, "Sounds good, but I forgot my golf clubs." No one laughs.

With Janesville just ahead, the rafts are again at full speed, even making a wake in the water. As they approach a sharp bend in the river they can see the park grounds sticking out into the bend. The three rafts easily slide up onto the grassy bank, just as the sun sets.

"Look, up on the golf course! Look! Hello? Hello?" Aaron yells.

Three people, two women and a man, are camped up on a green and are looking their way. Joel and Matt run to meet them before it gets too dark.

"Hello?" "Hello?" "Who're you? What're you doing?"

"We are on our way to Israel. We just need a ride. Heh-heh." the man says. "Actually we were expecting you, but not until tomorrow. Yoni told us to meet you here, and not to miss you. So we came early."

Matt says, "Although this looks like a pretty comfortable place, why don't you move your camp down closer to our rafts and join us for supper?"

"Much obliged!" the man says. "Shall we join them?"

"Of course!" the two women say.

Al, Linda, and Betty pick up their sleeping mats and supply bags and walk down to the shoreline where they enthusiastically introduce themselves to the others. Their story is somewhat similar, but nevertheless unique.

The teams now have their act together. Ian and Matt are off for wood, while Ellis and Miriam start a small fire on the beach. Others set up the camp. Teah, Rivka, and Raquel begin setting out the evening meal which begins with the traditional blessings and ends with the newly learned after meal thanksgiving blessing.

"Well," Isaac says, "I guess it is good that we built big rafts. Looks like we will be picking up passengers, I mean brothers and sisters, along the way. And don't misunderstand me, for that privilege I am very thankful. I'm sorry, I must sound confused?"

"Yes," says Teah, "he's had a long day."

Tonight, there is less talk and mingling. Everyone is exhausted, after what has been the most strenuous physical day so far. Everyone drifts off to sleep early.

CHAPTER 8

CHAOS

WISCONSIN – DAY 9

The thirty-six rafters are up and getting underway early, as the sun rises. After a quick breakfast, the rafts are packed and ready to go.

Joel says, "I grew up around here. As we head south today, we'll be going through some heavily industrialized cities like Janesville, Beloit and Rockford. I don't know quite what to expect, but I'll bet there'll be burned-out buildings, downed power lines, collapsed bridges, abandoned vehicles and boats and a lot of debris. You know, maybe there'll be some more survivors. So, let's keep close together and keep a sharp eye out."

"Agreed," say the rest in harmony.

Finally, all these heavy, water-laden craft push off into the rapidly-flowing Rock River current. Joel's raft leads the way, followed by Matt's then Isaac's.

The Rock River runs fast now, going south through Janesville. Indeed, as Joel has expected, there are a lot of burned-out buildings, warehouses, and downed high-tension power lines. There are abandoned, burned-out vehicles and trucks on the roads running parallel to the river. The river takes a straight course through the heavily industrialized banks. Sunken private boats litter the shoreline. But there is no sign of life and no sign of additional passengers to pick up. Ian continues to scan the landscape with his sniper scope. No signs of life.

The rafts move beyond the southern outskirts of Janesville and toward the burned-out dairy farms south of the city.

Isaac goes over and sits beside Teah, "Want to know anything about Janesville?"

"Sure, why not," Teah replies softly.

Isaac is pleased to have such a receptive audience and eagerly starts his story.

Janesville was settled like much of southeastern Wisconsin was during the Second Great Awakening. Very conservative and very religious farmers, engineers, and professionals from the northeast were coming west to escape the cold, frozen farms and villages of upper New York and New England. That was during the Little Ice Age, just after 1816. Those New England Congregationalists tried to impose their strict religious laws in their developing western settlements, but incoming Germans from proto-socialist and secular backgrounds fought back. The Catholics resisted too. I guess that battle's been continuing for way over two hundred years.

Among the most interesting of the groups who moved here were the Seventh-Day Baptists. They had European origins way back to the early 1500s. They later immigrated to New England in the 1660s, but had to seek refuge in Rhode Island. You know, that was the only colony with a right to freedom of religion, so they built a Seventh Day Baptist church in Newport. It's still there. Well, they prospered and spread throughout New England. At one time they were among the largest denominations in New England. They had and still have good ties with the Jews. In fact, their first church is located beside the first synagogue in New England, also in Newport.

But with the coming of that Little Ice Age, many of their members also moved west. In fact, so many relocated that the church decided to move its headquarters to right here in Janesville, in the Wisconsin territory. Around 1840, they built their first church here and then a few years later a new stone church was constructed that has been in use, last I heard, up until all this devastation. Nearby, in Milton, they founded the oldest college in Wisconsin, which was still open and active up into the late twentieth century.

Janesville became an industrial power-house, building cars and tractors for General Motors, and it was the home of the Parker Pen Company, manufacturing famous fountain pens that were in demand world-wide. Special editions of Parker pens were usually selected to sign important government documents, including treaties during and after World War II. General MacArthur used them to sign the surrender of the Japanese on the battleship Missouri in 1945.

All that industry, however, left Janesville due to the tight labor union control, back in the 1980s. The population declined afterward.

But now, all that is no more. I hope we'll see this town rebuilt one day.

Teah sighs, "Me too."

She grasps Isaac's hand and leans on his shoulder. Isaac wants to put his arm around her, but is shy and resists. He just holds on to her hand.

The rafts are making great time as the river is straight and the current is swift. There have been no river blockages and no disruptions. Within two hours they are coming into the outskirts of Beloit.

As Beloit comes into view, Ian stares thru his scope. "Guys, you'd better see this!"

Each person on Isaac's raft uses the scope to get a better view of what's ahead. Isaac shouts out to the other rafts who are in the lead.

"Joel, Matt, hold up, let's talk!"

The two forward rafts slow down and come along either side of Isaac's raft. The teams take turns looking through the scope.

"What do you make of this?" Matt asks Joel.

"It appears to be an industrial construction crane that has fallen across the bridge and over the river," Joel replies.

Matt says, "Dammit man. Anyone could see that. Look along the port side shoreline. In the distance there is a Gia Union Command Center!"

"Yeah, I see it now. It seems to have a big hole blown into it, unlike anything we've seen so far. And look around the Center . . . there are burned-out tanks. There's been a battle there. Maybe it was some type of a rebellion."

Ian joins the discussion. "A rebellion seems unrealistic to me, but havin' a Command Center in Beloit is not unexpected. Hell, it's a major crossing of north-south and east-west interstate roads and several rail connections, as well as the Rock River waterway. It would've controlled commerce for this whole vast industrial and agricultural region. But, why in hell there'd be a battle there is a mystery to me. Agree, we need to get a closer look."

Joel, Isaac, and Matt confer and discuss the observations with the others. All agree to advance cautiously. The rafts cut their speed by punting to the reverse, keeping the rafts moving more slowly than the current. They approach the broken bridge with the industrial crane on top.

Joel whispers, "It seems to me that the bridge is still intact. So are the supports for the collapsed industrial crane. It looks as if there is sufficient clearance to go under the bridge. It appears safe. Agree?"

"Yes," say Isaac and Matt.

And under it they drift. Beyond, the Gia Union Command Center is now clearly visible. As they move closer, scenes of destruction and of a great battle are becoming more evident. Almost all the buildings on the shoreline are not just burned or abandoned, which has so far been the norm, these buildings have also been

blasted from the sides and tops. Craters are all around, in the yards, the plazas and in the roads.

Joel says, "Holy crap! Looks like there's been one bastard of a battle. There are dozens of burned-out tanks with turrets open and treads broken, all pointing away from the Command Center. It appears that they have been protecting the Center."

Miriam breaks into a loud whisper as she points to the sky. "Look at those birds circling over the Command Center, I think they are vultures!"

All look up and are speechless.

Miriam says, "In all the destruction we have seen so far, there has been no vultures and never any sign of bodies. I think something's really different here."

Ellis suggests that they should pull ashore and investigate this mysterious mess. "I know we're not the Starship Enterprise, and it's certainly not our mission to explore, but there could be survivors. Hell, at a minimum, maybe tools and supplies we can salvage."

At first the teams do not treat the idea with favor. But after discussing this for a few minutes, the consensus is to pull ashore and reconnoiter. So, with a sense of trepidation, the three rafts pull over to a concrete boat dock and tie down to the marine cleats.

Everyone steps off the rafts cautiously and begins to wander around at random, but most amble toward the Command Center. The Center has a hundred-foot hole blasted through what appears to have been the front entrance. There is a deep blast crater just in front of the Center.

As she approaches the Center, Raquel quickly covers her nose to constrain the disgusting odor.

"Whew-ee! Smell that nauseating stench coming from the building and the tanks and personnel carriers. Oh my! I see pieces of bodies . . . arms, legs, heads, hands and guts."

Over to the right a dozen turkey vultures are feeding on a carcass. More vultures circle overhead.

Aaron points to the encircling areas. "The surrounding buildings have much smaller blast holes, maybe from artillery or tank cannon shelling. I can see dozens of shattered trucks and vehicles in the distance and, damn . . . many of them appear to have bodies in them. Look over there, there are the ruins of four helicopters scattered among those vehicles. Whoa! There's the friggin' wreckage of a crashed F35 attack jet on the horizon. Look there! It's an unexploded cruise missile on the plaza. Everyone—Keep clear of it!"

Raquel, who has a nursing degree, says, "All this appears to have been very recent, perhaps within the last three to four days since the bodies are in a very early stage of decomposition. There are some maggots in some, but not in others. Well, we now know that some vultures have survived, but I think these birds are new arrivals here as well."

Raquel and Aaron move toward the blasted open door of an armored building in the periphery. Aaron slowly sticks his head into the building, before stepping inside. Raquel is right behind.

"Help!" "Can you help us?" A weak, crackling voice is coming from inside the darkness. Aaron and Raquel step back. Raquel runs to get reinforcements from the others. The full team runs over to the building.

Not seeing anyone, but hearing the voice continue to whisper, "Help," Aaron moves further into the building and inquires, "Who are you? Where are you? Are you alone?"

Isaac, Ellis and Ian quickly join Aaron and instinctively react into assault formation. They move into the building as if on a special operations mission, but without any weapons. The voice is identified to be coming from the third door to the right. Moving along with backs to the wall, Ellis opens the inward-swinging door with his right foot. By now their eyes have somewhat accommodated to the dark.

They can see a face. No, it is several faces. But the voice is coming from the one in the middle. All the bodies are covered with collapsed building debris, and they appear to be trapped.

Aaron loudly repeats, "Who're you? What're you doing here?"

Then he adds, "Are you Gia or rebel?"

Ellis whispered to Ian, "That's insightful, huh?"

"No, not Gia . . . rebel," the weak reply comes back.

"Okay," Aaron says, looking at the others, who nod. "We'll help you as we can."

"Thank you sweet Jesus," is the feeble response.

Six others of the team come to assist. Raquel insists she supervise the extraction, since she knows something about assessing for fractures and injuries. All agree.

For over an hour the piles of debris are removed one piece at a time, as they continue to uncover and check the condition of the rebels. It is a slow process getting to the first two. Taking reed mats from the raft for stretchers, Ellis rolls the first rebel onto a mat. The four men slide the mat to outside on the plaza. The same process is repeated for the next one.

There are four others buried even deeper, but the methodical extraction continues until all have been removed and placed onto more mats. All are lined up outside the building on the pavement.

Raquel, who has been monitoring the removals and examining the rebels, reports, "These guys are all still alive, two are conscious. Only one is able to speak, but all are severely dehydrated and in shock. I can't determine the status for any fractures or head or spinal injuries yet.

"Linda, Betty, Teah, Miriam . . . go to the rafts and bring back some fresh water.

"Guys, let's move them gently back across the plaza, closer to the vehicles where there may be other survivors."

Except for the idea of searching the other buildings, the team swings into action.

Teah returns quickly from the raft with six bottles of water, but she has some concerns.

"Let's help these guys as best we can. But we don't know what has happened here. Are we sure we're helping friends or foe? If you were an injured Gia Trooper, how would you answer the question by a non-uniformed stranger, 'Gia or rebel?' I think lying would not be beneath any Trooper."

Raquel says, "And has anyone noticed that they all have on Gia Trooper uniforms? Did anyone notice they all appear to have The Marck in their right hand?"

Everyone pauses. No, no one but Raquel and Teah has taken time to notice. They have just been focusing on rescuing fellow human beings.

Isaac leans down and gives the first Trooper a sip of water. "Sip slowly, but keep sipping."

Raquel does the same to the other conscious Trooper. Linda, Miriam, Rivka, and Joel began to pour small amounts of water into the lips and onto the tongues of the unconscious ones. This continues for over an hour. One of the others is waking up. He is looking about, seeming to be confused, agitated, and frightened.

When the vocal Trooper has improved his hydration, he asks for food. Linda and Betty go to the rafts and bring the bread Yoni had given them. A small piece is given to the first Trooper. The other two conscious ones are also offered bread.

One takes it and the other rejects it, shaking his head, still frightened.

Joel calls the teams into a huddle. "It's now approaching two o'clock. We have four more hours of daylight. We'd planned to be through Rockford by now and a good way toward Rock Island by nightfall and we don't know who else Yoni might have arranged for us to pick up. Now, if we take time to search the periphery buildings, not to say the vehicles and the Command Center, it'll take us the rest of the day. And what if we find more, we have no medical supplies, limited food and space. And, as Teah observes, we don't know whom we are dealing with, or what went on here."

Isaac looks up from the Trooper he is helping. "You think we should go on and leave them?

"No, I'm not saying that we should just leave them alone, but we need to have a talk and an agreement on what we should do. And we need to decide as soon as possible."

Linda, Miriam, Raquel, Rivka, Isaac, and Joel continue to give sips of water to the rebels, along with bread to two conscious ones, while the discussion goes on.

Finally, Isaac leans down over the first Trooper and asks, "Feeling better? Can you now talk? What's your name? Where are you from? Do you know where you are?"

Raquel leans over and says, "Not so fast Isaac, give him a chance to tell his story."

She gently asks him, "Okay, now tell us how you got here? Can you do that?"

He nods, and slowly and softly begins. "I understand what you are asking and I heard your discussions and your concerns. But I do not know who you are either, so please believe me. We are just some God-fearing sinners who tried to help Jesus. We thought the world had come to an end, and we were all going to die. So, it seemed to us the one thing we could do was to take out a Command Center and do as much destruction as possible."

Isaac asks, "How many were with you? And who are you?

"There were almost a hundred of us, all from Seventh-Day Baptist churches near Janesville and Milton. So much has happened, it's hard to know where to start, but the bottom line is that we were tricked into taking The Marck. The Gia Union was very persuasive, not demanding, just persuasive. They recruited us as a platoon. We agreed only when we were promised our own unique unit. We could continue our own form of worship. We could serve the Union in non-combat roles. We would be given good salaries and we would always be attached to the Beloit Command Center, where we would be near our families, not far away in Milton. Our platoon continued to meet together on the Sabbath and to worship, but we soon realized our mistake when the members of our congregation, who did not take The Marck, disappeared and we never saw them again.

Raquel asks, "Is that when you decided to resist?

"We really didn't know what to do or how to do it. But after the Supreme Commander proclaimed himself to be The God, our unit became more and more isolated, and we were distrusted by the local Commanders. They were not concerned about which day we worshiped on, but increasingly insisted we direct at least some of our worship toward the Supreme Commander, Gog. We just couldn't do that and continued to be faithful to our religious convictions.

Teah says, "This looks like it was an unbelievable battle. Especially with such a small number of dissenters."

"We realized our time was numbered, in all too many ways. So we begin to stash away weapons and explosives. Years ago that would have been impossible, but after the last firestorm and electromagnetic pulse, communications were disrupted. The Marck positioning system, to keep up with our movements, was severed. Buying and selling could no longer be managed by the Command. Communications and computers were out and bartering was the order of the day. With Gia Troopers, we could arrange to barter goods we got from our community in exchange for stuff to be used as weapons, ammo and all.

"Believe me. We all are long-term, serious students of the Bible. We knew what was going on. We knew that the plagues were from God, to punish Gog and his government and those who worshiped him. And we knew his overthrow was imminent. So we continued to amass weapons and explosives, and even got access to armored carriers and a couple of 'copters. With all the confusion over the last month, I must say that no one seemed to be paying attention to us anymore. It was becoming every man for himself. Many Troopers took off their uniforms and went AWOL. Still the Command could not restore effective control.

"Now, too late, we all realized that since we'd taken The Marck of the beast, we'd have no hope, only certain destruction awaited us. So we had nothing to lose. We developed our final plan."

"Then we heard the Sound of the Trumpet and we all saw the hologram-scroll of Jesus, the Son of God, returning to Jerusalem. We knew the overthrow of the beast and his government was about to happen. Our souls were lost for having taken The Marck of the beast, but we were still in a position to help in the overthrow. We had weapons and explosives and were motivated with a sense of payback.

Isaac says, "A Command Center like this usually has four to five thousand troopers. I really want to hear how less than a hundred men created this much devastation in face of that kind of opposition.

The rebel Trooper says, "In those days after the hologram-scroll there was total disorder in the Command. Then, a couple days later a signal got through. It was an order for all of us to move-out the next morning. We were going to Har Megiddo. And we Bible believers knew what that meant. We also knew that it was time to take action."

Raquel interrupts, "Here take another sip of water and a bite of bread, and get your breath. But please continue."

The whole team has been listening to this stirring account. All are now tightly squatting down, leaning over to hear more.

He continues, "That night in the Command Center, the Commander and leaders were drinking heavily while they and all the Troopers were busy packing and getting their gear and affairs together. While they were distracted, we drove a truck packed with high explosives, including two tons of fertilizer nitrates, into the compound and parked it near the front door of the Center.

"The guards just thought it was another transport vehicle. Security was their lowest priority right then. Knowing there would be no Troopers in the bordering armored buildings, we stationed ourselves in them with our mortars, RPGs and antitank weapons. We also positioned night-vision-equipped snipers on the roofs."

The four Marines looked at each other and silently nodded their approval of such a brilliant plan.

"At midnight, we remotely detonated the truck! What an explosion! A blinding white and blue flash of light filled the entire city and the shock wave knocked us all back against the walls. Every vehicle in the plaza was overturned and many went up in blazes. A large illuminated cloud shot up into the sky, lit up by the fires and explosions. After that, we watched the Center carefully using our night vision goggles, but very, very few survivors were seen leaving the building. The few that did, our snipers took out.

"But, as we expected, a counter attack soon came from behind the building, a column of tanks and overhead attack 'copters. And a jet aircraft zoomed overhead. The stupid tanks formed a ring around the Command Center, as if there was still anything there to protect. Our anti-tank missiles took them out, one at a time. Many crews opened their turrets and poured out, only to be picked-off by our snipers. Heck, our snipers had their greatest night ever. An attack 'copter, hammering me with dual M60D cannon fire, swept over my building. As he returned, I stepped outside and fired my shoulder-mounted Stinger Z . . . and out he goes. Other 'copters began to swing into motion. We even had a couple of them ourselves. Yep, we took out one of their jets too.

"We fought a good fight. We took out some really bad guys. I pray we helped out Jesus some too.

"But then it was only a matter of time. There were more tanks, more 'copters, more jets, and they also had snipers. Once they had us pinned down, we knew it was the end. We held out until early morning when we took our PR-15s and retreated into an inner room. We were expecting a room-by-room Special Ops type assault, but then the outer wall of our building was hit by . . . by . . . by I don't know what. And things simply went dark. Explosions from the outside stopped. It was just dark and we were all covered by debris, unable to move. No Troopers came. I guess they thought their job had been done. Besides they were all still under a lot of pressure to move out to Har Megiddo.

"We were sure we were going to die. We all expected our next conscious moments would be standing before Jesus' throne, being condemned for taking The Marck and being sent directly to the fiery depths of Hell. I drifted off, lost consciousness, in terror.

"Then I awoke. I think maybe one or two days ago. I was unable to move, barely able to talk and pinned in with no food or water. My friends all seemed to be unconscious or dead. I drifted in and out of consciousness. Then I heard your voices outside. I thought that maybe some of our folks did make it through, or maybe it was the Gia Special Ops Troops finally making a sweep. What did I have to lose. I called for help. And, thank you Jesus, thank you, thank you Jesus, here we are."

He pauses to get his breathe, and finally, exhausted, but with a weak smile, he asks, "Now, what's your story?"

No one is able to immediately respond. Some cry, some dip their heads in prayer. Al, Linda and Betty are looking up, hands raised upward to the east mumbling prayers of praise.

Teah says, "*Me chah mocha, ba'al g'vurote!* (Who is like You, O Lord of might?)"

Most stand or squat in protracted silence.

Finally, Ellis, as is not unusual, breaks the silence, "Okay guys, that answers our question. Now we're going to have to kick-ass to get these armored buildings searched before sundown. And it looks like we will be spending the night here. Any questions?"

Aaron turns to the first rebel and asks, "It would help a lot if we knew exactly how many men were in your unit. Do you know?"

He responds with confidence, "Going into battle, we had ninety-seven."

The group organizes into teams of four. They spread out in four directions for sweeps of the twenty peripheral buildings. Three remain with the rebels, continuing to give sips of water and bread.

When a body is identified, Raquel is called over to determine if he is alive or not, if it is not obvious. Teah is organizing an ongoing count of the team findings and keeps record of the buildings surveyed, in order.

"At the start we know there were ninety-seven rebels. Two in 'copters, which we have searched, yielding two bodies, both dead. There are six from the first rescue and eighty-nine more are to be accounted for," she reports.

Matt yells, "Six bodies over here ... appear dead. Raquel, come and confirm."

Then Isaac, "Two here . . . definitely dead."

Ian, "Four here . . . dead. Teah, keeping up with all this?"

"Yes sir!"

And the body count goes on for an hour.

Then Aaron yells, "Here, one's alive and maybe three more. Raquel, get over here!"

Raquel rushes across the plaza to see. She quickly confirms, "Yes, all four are alive. Let's get them some water and move them outside the building."

Ellis calls out, "Four here . . . all dead."

And so the building survey, search, and body counts continue over the next two hours.

Ellis says, "Okay, we have swept all twenty buildings and two 'copter ruins. So, what is the count, Teah?"

"Including the six originally rescued, we now have fourteen alive. There were two bodies in the rebel 'copters, and a body count from the twenty buildings of seventy-nine. Total count is ninety-five. There are two missing."

Ellis asks, "Should we sweep the buildings again, or did two men escape, or maybe didn't participate?"

The quick consensus is, as dreadful as it is, there must be a second sweep through the ruins. The teams divide in thirds. Team one helps move the additional survivors to the camp and provide water and food. Raquel stays with them to continually reassess their situations.

Teams two and three run through the second sweep of the twenty buildings as the evening shadows lengthen over the Gia Command Plaza. No further survivors and no more bodies are found.

Al says, "Since we'll be spending the night here, maybe we can do a third sweep in the morning in better light."

Trooper number one, now to be called Rebel number one, is stronger and able to sit up. His vocalization is improving. He is helped to his feet to check the other survivors. Many of them are gaining consciousness and he chats with each one that he can. He wants to look over the other bodies that had been found, but they are outside the buildings, a long walk, and darkness is imminent.

"Maybe in the morning," Raquel says.

It is time for dinner, but no one is hungry. Then there is that foul odor coming from the Command Center. Everyone feels dirty and defiled. It is decided to forgo a formal meal tonight. Berries and bread are set out for anyone wanting to snack.

Teah says, "I think I need to bathe in the river, a really good soaking." Others agree.

"Ladies first. When you all are out, whistle, and it will be our turn," Isaac suggests.

The water is chilly, but refreshing, and while in the water they wash and rinse their clothes also. All this takes two hours. It is fully dark when everyone gathers again around the inviting fire, Ellis and Matt have built, to warm and to dry.

There is a shortage of reed sleeping mats, as fourteen are delegated for the injured rebels. But everyone is very tired and barely miss the sleeping pads. The teams divide into four watches to monitor the injured overnight, providing drink and bread as needed. Ian, on the first watch team, leans forward and reflects, "This is quite a big change in watch duty."

The watch is now not for warnings of Gia Trooper attacks, but to care for injured Baptist Gia Union Troopers.

CHAPTER 9

ARMAGEDDON - GOG RETURNS

JERUSALEM – DAY 10

It is very dark. The waxing moon has already set. It is 3:00 a.m. Quickly and all together, responding to the "move out" signal from the Supreme Commander, all Gia ground and mobile forces move south toward Jerusalem. Mission Retribution is underway. Within an hour, advanced units of the Gia Union Troops have landed on the mountains surrounding Jerusalem, without any resistance. The buildup continues. The main forces are massing on the high northern plateau, again without resistance.

As the sun rises over the crest of the Mount of Olives, millions of Troopers have Jerusalem surrounded, cut off from all sides. More Troops are surging in from the north. An Operations Center for Mission Retribution has been established on Mount Scopus.

Stan says, "Gary, this is one damn great location! From here the battle can be directed, just like they told us that Vespasian and Titus did two thousand years ago, and Nebuchadnezzar six hundred years before that. This is remarkable! Hell, you can just call me Titus and I'll call you Nebuchadnezzar."

"It's fricking fantastic! I'm so excited. We are going to do in one day, what took Titus and Nebuchadnezzar months!"

Gog himself is commanding from the Mount Scopus Ops Center. His Prime Minister Dajjal is with him. His trusted guards, Gary and Stan, stand at his side watching, advising, and monitoring the Commander's personal laser reflection shield. And HaSatan (Satan, now choosing to appear visibly as a fierce Red Dragon) observes all this from further away to the east, high above the Judean desert. Gia Union Recon Sky Pods hover overhead at 20,000 feet, recording and transmitting visuals and targeting data for the battle below.

It is action time, as the first sunbeams shine on the Temple Mount. The command is finally ordered.

"Attack!"

Gary and Stan give a loud spontaneous, "Ooo rah," and raise their light-swords.

Twenty cruise missiles and ten stealth bombers initiate the attack, their arrival coordinated from bases in surrounding countries (Egypt, Jordan and Turkey) and from the naval units off the coast and from the Red Sea.

Then moving forward, tank formations and armored personnel carriers tighten the circle from the north. Thousands of individual armored vehicles follow them. The attack is sudden and swift. It is intended to be a blitzkrieg.

Within minutes of the attack order, the missiles and bombers arrive over Jerusalem.

Cruise missile warheads explode in a rapid timed-sequence, directly over the Old City, the Temple Mount, the City of David and pre-targeted business and government centers along Joppa Road. Loud booms and aerial shrieks are deafening. The blast waves shake the ground. Seconds later, twenty three-thousand-pound bunker-buster bombs from the stealth bombers explode on target, the same sites. A double whammy for each of the twenty selected sites.

Dense columns of black, brown, and gray smoke arise and coalesce, covering the entire seven mountains of Jerusalem and her valleys. Nothing definite can be seen from the Mount Scopus Ops Center. Views from the Recon Sky Pods show only dark rolling clouds of smoke. Stan looks at Gary and nods.

Gary looks at Gog and Prime Minister Dajjal and makes his first assessment. "Great! Looks like we got 'em! Caught 'em with their pants down!"

Everyone in the Ops Center raises their fists and cheers, "Long live Gog. Long live Gog. Long live Gog!"

The celebration starts and a first bottle of champagne is quickly opened. A southerly wind begins to move across the city, gradually dissipating the dark, smoky cover.

When enough smoke has cleared, visuals from the Recon Sky Pods come up on the large high definition screens at the Ops Center. And now one can also directly see from Mount Scopus. The celebration tapers off, then stops. Within minutes, all are speechless.

Stan is first to get out a comment. He yells, "Holy Shit! There's no fuckin' damage! The city appears to be totally unaffected . . . including the Temple reconstructions. Not a single target damaged. Damn, it can't be! There is no effect on the city anywhere! It looks like there's some kind of invisible shield covering the whole city."

Gog asks Gary, "Do our recon data really show that? Really? Could this be some kind of optical illusion? Is it? Can we verify no damage? Yeshua would just love to fool us, to play with our minds. If he survived, that is."

Gary and Stan and a dozen technicians scan the area with visual and IR scopes and analyze the Recon Sky Pod data.

"No sir, I'm afraid to say. All the data confirms what we see-- there is no damage," Stan reports. "Now there's gotta be some kind of energy shield. One we cannot see and our sensors cannot detect. But it's gotta to be there, sir."

Gary thinking quickly, analyzing the situation, interrupts and speaks directly to Gog, "Sir, we did this initial attack with conventional precision warheads, to avoid nuking the place, to limit radiation and such, but . . . but, I did plan an option B. And we must have drained a lot of power from their shield. I recommend that before their shield can regenerate we should immediately send a second wave, this time with tactical nuclear warheads. They are ready to go, I have them in the air already, awaiting your command."

Gog replies instantly. "That makes sense, Colonel, I concur. Make it so."

Within three minutes, at higher altitudes, a second wave of bombers with tactical nuclear-tipped missiles suddenly appear and deliver their goods. Suddenly, one by one, seconds apart, there are two dozen direct hits on the city and suburbs. A rapid series of blast-pressure waves undulate through the valleys and hills, followed by earthquake-like ripples traveling through the ground. This series of explosions leave a chain of twenty-four mushroom clouds, each swirling counterclockwise and rapidly ascending to twelve-thousand feet, before coalescing.

This time within the Gia Ops Command on Mount Scopus, there is more restraint. All watch intensely as the towering mushroom clouds combine and swirl and spread out overhead. All watch anxiously and await data from the Recon Sky Pods.

After ten minutes, a southerly wind reappears and again gradually moves over the landscape. Slowly, clearings in the mushroom clouds begin to appear.

Stan focuses his infrared scope. "No way! No way in hell! This just can't be. I see intact buildings. I even see people. They're standing there, praying on the Temple Mount ... and there's no harm to the city ... or the people! Crap! That friggin' invisible shield is holding. There's no penetration."

The southerly wind moves across the hills and into the valleys. The smoke and radiation is swept northward into the path of the approaching 'copters, tanks and armored carriers and over the Ops Center on Mount Scopus. Momentarily the sky is dark again on Mount Scopus. But visuals from the Recon Sky Pods clearly reveal the battlefield.

Without consulting Gary, Stan or any of his Counsel, Gog steps forward and screams his command. "Continue the attack! We will overwhelm them, the ol' fashion way."

Stan watches as more Troops move down the hills, filling the Kidron Valley to the east and the Hinnom Valley to the south and west. Others come from the north and northwest.

Stan reports, "They are beginning their ascent upward toward the Old City walls. Nine thousand of our elite Gia

Special Op Troopers are now fast-roping from hovering 'copters onto the walls and gates of the old city."

As the assault on the walls and gates start, it is now nine in the morning, at the time of the morning offering and the time of the morning prayers, which today had started an hour earlier. Yeshua is among the people east of the Holy Place. He steps out from among the crowd, walks up to the bronze altar, the o'lah altar, and lifts his hands. He turns and looks in every direction, observing the armies in every direction, and the Troopers on the gates. He now faces west, toward the Holy Place. And He leads the people to proclaim in unison,

> "Hallelujah.
> Yehovah our God, the Almighty God reigns!
> Yehovah is our God. He is our refuge. He is our high tower.
> He is our shield. He is the Shield of Abraham.
> Yehovah is our God, Yehovah is a God of war!
> And now has come the Salvation and the Power of the Kingdom of God.
> Blessed is He, who comes in the name of Yehovah."

Then the crowds pause and listen in silence while Yeshua concludes alone, but with a voice that echoes through the city and valleys,

> "Blessed are You, Father, the God of Abraham, the God of Isaac, the God of Jacob!
> Blessed are You, Father, the Shield of Abraham.
> By the Authority of Your Messiah,
> Save Your people, Your city, Your land.
> Baruch HaShem!"

Abruptly, a bright, blinding-white beam rises from the site of the Holy of Holies. The beam immediately shines upward, then widens as it expands downward and rotates around the horizons. The beam is accompanied by ear-piercing high pitches, sounds and deep vibrations. A great earthquake begins with a sudden upward thrust. Then the shaking changes, horizontally, with great north-south rocking movements. It is a 9.7 quake that lasts for seven minutes.

There is surprise and confusion among the ranks of the Troopers on the walls and gates, as well as within the Command Center. The high definition battle communication monitors have gone blank.

"What the hell's happening?" Gary shouts.

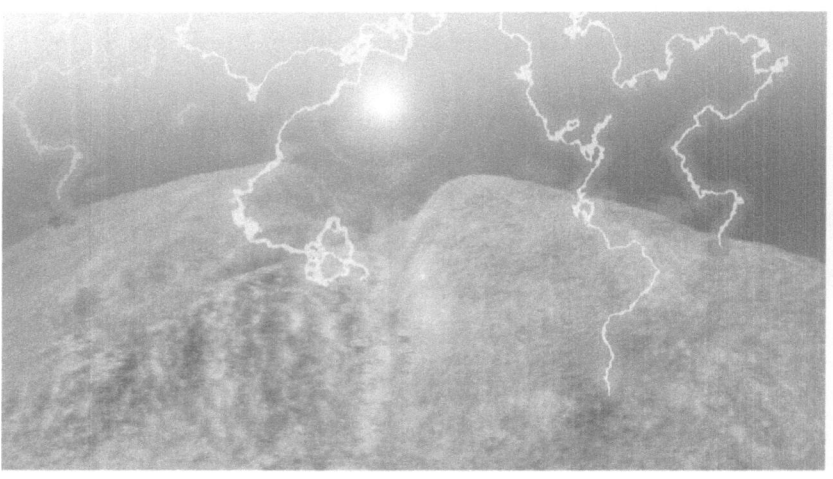

Then with a deafening thunder, the Mount of Olives begins to split. One half moves northward, the second half to the south. A torrent of water begins to gush out from the rocky crevice over the Gihon Spring. It gushes into the Kidron Valley and into the opening, down through the new deep valley that has been formed through the Mount of Olives, now split in two. Hundreds of tanks, personnel carriers and thousands of Troopers fall into that new valley. The helicopters hovering over the walls lose power and fall, rolling as fireballs into the valleys below.

At the time of the flash, a massive electromagnetic pulse enveloped the region for eighty miles in all directions.

Gog yells, "My shield! It's down. Get it back online! Now!"

Gary and Stan are speechless. They quickly assess the situation.

Stan shouts, "All electronics, including all our shields, our lasers and beam weapons are fried. My light sword won't even turn on. Our digital and optical communication systems are fried too. There's no communication from the Sky Pods, the aircraft, the ground units. And, oh shit! My repair instruments ain't working either."

Gog makes a threatening lunge at Stan, demanding immediate protection.

Gary and Stan, without delay, command the Guards to move down the mountain in a three deep semicircle, providing a perimeter fortification for Gog and Minister Dajjal. But they have no weapons, other than their ceremonial swords, bayonets, and knives.

Transformer and circuit explosions are starting fires everywhere. The laser and electron-beam weapons the Troopers carry are overheating and exploding. The same is true in the airplanes and attack 'copters overhead. They have lost navigation and power and fall out of formation. Planes and 'copters begin to crash everywhere, rolling down the steep hills into the valleys, crushing thousands of Troopers on the slopes and massed in the valleys.

As millions of Gia Union Troopers experience and watch all these unexpected events, a sudden and paralyzing fear envelops them and their command structures. They are unable to launch their planned massive strikes on the city.

The planes, the tanks, the artillery, and the rockets have lost their computer-crystal links and their cyber-integration controls. Nothing works.

On the city walls and gates, the Special Ops Troopers, who have made it that far, fall off the walls and high gates, crushing themselves and those below. Others turn back, repelling down the walls, and retreating down the steep valley slopes. There they encounter other confused units now pushing against those poised to move upward. On the steep slopes and down into the valleys, soldiers and Troopers begin to push and shove each other. Fighting breaks out among the many nationalities and mercenary units. Their laser and beam weapons are useless. They have no communication system. The fighting has to be the old-fashioned way, hand-to-hand combat with fists, knives, swords, and bayonets. Blood begins to flow ... into puddles at first. It's every man for himself, and against everyone else.

As the confusion and fighting spreads, the puddles of blood and body parts increase and begin to coalesce into streams of blood. The streams merge and flow southward down through the Kidron Valley spilling over into the Gihon Spring. The torrents of blood pollute the stream and flow through Hezekiah's tunnel and out into the Pool of Siloam. The blood mixture gushes into the newly formed channel toward the new Mount of Olives valley.

This polluted pool of blood and water is at some sites four-feet deep, as high as a horse's bridle.

The fighting no longer has any directional orientation, or organization. Most randomly try to push their way downward away from the city itself, into the valleys and then back up the adjacent slopes. Throngs of bloody, wounded Troopers begin to push up the slopes of Mouth Scopus, engaging the perimeter set up by Gary, Stan and Gog's other Personal Guards. The Guards fight valiantly, but fall themselves one by one. Gary orders a retreat to form a smaller perimeter as dozens more fall.

Suddenly Gary experiences a sharp pain in his right chest as a bayonet penetrates his thick leather vest. He catches Stan's eye briefly, before losing consciousness.

Stan and four others withdraw toward the mountain top where Gog and Minister Dajjal have retreated and are watching in fear and horror. Stan has never seen such expressions on their usually self-confident faces. As Stan looks back over his shoulder toward Gog, he suddenly experiences a sharp, deep pain in his left flank and chest. He leans over, and then another sharp pain on his right side.

Lying curled up on the ground, Stan stares upward at the face of a terrified Gog. This fearful image lingers for a few seconds, then gradually becomes blurry and dimmer, and then . . . fades away.

Throughout the day the fighting continues and spreads outward from Jerusalem and northward until the vast armies massed against Jerusalem have destroyed themselves, extending all the way back into the Jezreel valley. A stealth bomber has even crashed into the Gia Command Center atop Har Meggido. A 90-mile stretch of Israel, from Jezreel to Jerusalem is covered with mangled bodies, equipment, burned out tanks, armored carriers, crashed planes and helicopters and Sky Pods. Piles of bodies, limbs, heads, and puddles of blood liter the landscape. It is total chaos, a total carnage. Not one is left . . . the enlisted, the officers, the privates and captains, the generals, the mercenaries and even the Personal Guards. The Supreme Commander is watching in agony and terror as his mighty army destroys herself and its blood mixes with the new river coming from the Temple's rocky foundation, flowing into the new valley. All Gog's advanced technology is useless.

Now, only the panicked leader of the Gia Union and Dajjal, his Prime Minister, remain alive on the heights on Mt. Scopus.

From this same site two millennia earlier, Titus the Roman general had watched his long, but successful siege and destruction of Jerusalem, as had Nebuchadnezzar six hundred years before that. This time, however, the final king of the Final Roman Empire and of the Final Babylon, along with his companions, witness a very different outcome. For the first time in their lives, Gog and Dajjal begin to experience that sickening, sinking feeling of impending doom. They would run, but there is no place to run. They would hide, but there is no place to hide.

As evening draws on, at about the time of the evening sacrifice, Yeshua and two of his strong men, archangels, hover over these two beaten but still defiant mortals. They both continue to shake their fists and hurl curses at Yeshua.

Two archangels, Michael and Gabriel, descend behind the two miserable creatures and bind their legs in chains of light. They then lift them high into the air above Mount Scopus. The two captives dangle from the light-chains, heads down, as they begin a flight over the chaos below and across the city. They fly over the Mount of Olives, over a blood-stained Kidron Valley, across the new Olivet Valley, then turn to fly over the Temple mount. They make a final flyover of Jerusalem, for all to see clearly the sight, before moving over the southern crest of Mount Zion and to the southernmost ridge of the city, to the brink of the ridge of the Valley of Hinnom, better known to the world by its Greek transliteration, Gehenna, a representation of a valley of fire. Tragically, three thousand years before, this site was the place where perverted Israelites, as a sacrifice to the false god Molech, laid their young children into the arms of the Tophet in this valley. Flames below burned them alive, until their limp bodies fell through the Tophet's outstretched arms and down into the infernos of burning garbage below. The valley has remained a dump, with fires continuing to burn garbage and whatever is to be cast out and destroyed, up until modern times. The Valley of Hinnom, Gehenna, has been the symbol of hell-fire for ages. And, yes, in the new world of Messiah, fires in the Valley of Hinnom are to be continuingly burning fires, as a reminder of the foolishness of rebelling against the new King and His Righteous Government.

First, Michael drops the Gia Supreme Commander, Gog, into a great fire prepared for them in the valley. Gabriel follows suit by dropping Dajjal, Gog's Prime Minister and False Prophet into the valley of fire.

And so, the fate of the Gia Union Supreme Commander and his False Prophet is that of a burning garbage heap, as a testimony to the world for generations to come.

The events of this momentous day have been projected onto that great global sky hologram-scroll. And the whole world has seen these epic events – in real time action.

But there is more yet to see. Just as these two miserable creatures are being dropped into the Valley of Hinnom, one more important episode is underway. The other major player in these recent events, in fact for thousands of years, is the third part of that unholy trinity. It is HaSatan, known variously as the Satan, the Adversary, Lucifer, and the Dragon. This is the Devil. In person.

He has been hovering high in the southeastern sky over the Judean Desert while this "mother of all battles" was underway. One cannot image that he actually had any hopes of a final victory over Yeshua and of replacing him as world ruler, as he has been in such battles before. Now again, his newest attempt has just gone awry. His time of influence over mankind also is over finally, well, for a thousand years.

Sweeping onto this great sky hologram-scroll screen comes another great archangel, Raphael. He is visible as he swoops across the desert sky, light sword and chains in hand. HaSatan turns to flee, but is overtaken by Raphael. There is a fierce struggle between the two with each trying to get a position of advantage over the other. HaSatan strikes Raphael with his tail, then grasps at his wings. He attempts to twist his tail around Raphael's legs. He grasps at Raphael's sword arm. Raphael does a 360-degree head-over-heels flip, freeing his legs, and then maneuvering from behind, quickly grasps HaSatan's neck with his right arm.

HaSatan violently twists his body and pounds Raphael with his forearms and wings. Raphael then whips the chains he has in his left hand around HaSatan's neck and arms and then, with the edge of his sword, stretches the chains around HaSatan's tail and legs. HaSatan is still thrashing his wings and tail, but now, cannot escape. The chains are unyielding under the tail and across the wings. With these hyper-entangled-quantum-energy chains, HaSatan is now completely bound. Raphael now drags him across the sky, high over Jerusalem. Since the hologram-scroll is three dimensional, it is easy for all to see the menacing and mysterious orifice of a galactic-quadrant wormhole opening up overhead. Raphael stretches backward and with all his might hurls the bound HaSatan into the newly opened wormhole orifice in the northern sky.

This wormhole stretches 25,000 light-years to where it empties into the massive black hole located at the center of the Milky Way galaxy. The hologram-scroll follows HaSatan's trans-warp slide thru the wormhole, and his disappearance beyond the event horizon into the galactic black hole.

The name Raphael means God has healed. He was given the honor of this task, to finally rid the world of HaSatan, the originator, the source and the personification of greed, envy, sin, war, evil and death. Raphael's long-awaited task is now completed.

Raphael hovers back toward the sky over Jerusalem, pauses for a few moments, high in the sky, as if posing for the cameras, and then stretches out his enormous wings and holds his hands up high in a very happy victory pose. Some say that he has been waiting for this for a very long time.

Now, the hologram-scroll gradually fades and eventually disappears. The sun is setting and the evening shadows fade over Jerusalem, the capital of the world.

Yeshua had summoned his Kadoshim from all over to be present during this day's events. They now join the four mighty archangels, Michael, Gabriel, Raphael and Uriel. They will dine together this evening in peace and harmony.

This day has been Yom Kippur, the Day of Atonement. It was a fast day, always ending with a big meal, a traditional breaking of The Fast. And so it does this time.

After that dinner with his Kadoshim and his angels, Yeshua stands and salutes his team.

"Good job, team! Good job!

He smiles and winks. "I couldn't have done it better myself,"

Then each member of the vast host raises his and her wine glass high with the right hand. All say in unison,

"To The Kingdom!
To The Kingdom!
To The Kingdom!"

CHAPTER 10

RAPIDS

WISCONSIN – DAY 10

Everyone is awakened just after midnight by another hologram-scroll in the eastern sky. The battle for Jerusalem is underway, and the world has a front-row seat.

The Rock River team, with their new Seventh-Day Baptist guests, watch as Jerusalem is protected and the armies of Gog destroy each other. They watch as Michael and Gabriel bind Gog and his False Prophet-Prime Minister and cast them into the valley of Hinnom. They see Raphael bind HaSatan and hurl him into the wormhole leading to the giant black hole of the Milky Way. They all cheer loudly as Raphael struts with his victory pose. By the time all of this is over and the hologram-scroll is fading, the sun has already risen in Wisconsin.

Obviously, no one slept through all this. At daybreak, everyone is filled with enthusiasm and excitement by having witnessed the battle of Armageddon. Nevertheless, everyone is physically and emotionally tired from the previous day's stress. There is a constant buzz about the events that they have just seen.

All are not fully alert, but the injured rebels are doing better and most could witness last night's dramatic presentation. Interestingly, according to "Dr. Raquel," as everyone is beginning to call her, there are no broken limbs, no spinal injuries and no brain injuries. Just blast-related concussion amnesia and confusion.

"They will recover from this, I think," she says to the teams.

"Look, up in the sky!" Joel is the first to see Yoni approaching.

Yoni lands on the boat ramp beside the rafts and sits down on the big log on the port side of Isaac's raft.

"Looks like you all had a big day yesterday. And, of course, we all had a big night last night. We were expecting you to be well beyond Rockford, on the way to Rock Island by now. And looks like you have taken on some additional passengers? Did I tell you to find these men and to bring them along?"

There was a long silence, and he allowed the silence to linger before continuing.

"I did not tell you what to do here. I left it to you to discover, to decide and to act by your own volition. This was a test.

"And . . . you have passed! Looks like you can do this yourself. Good decisions, good compassion, good work. "Couldn't have done it better myself.

"Well, maybe a just a bit!" Smile. Wink.

He walks over to the fourteen men on the reed mats, reaches down to each one and lifts them by their left hands to standing positions. He touches each on his chest. Each man feels an electric tingling throughout his body. They are healed, instantly.

"Hallelujah, Praise be to Jesus!" they all say together, in a way you might expect of Baptists.

"Amen," Yoni replies.

Next he stretches out his right hand, grasping the other man's right hand. The hand radiates with a soft red glow, which then fades.

The first man says, "Have you neutralized The Marck?"

"It is gone." Yoni goes down the line, removing The Marck from every one.

The first Rebel speaks up and says, "Does this mean Jesus will not burn us in the fires of hell because we accepted The Marck?"

Yoni pauses, and replies, "Well, you can discuss that with him when you see him in Jerusalem." Smile.

Teah then reports to Yoni the body count and the fact that there are two missing rebels. Yoni says, "Got away! Don't worry, we'll pick them up along the way."

Everyone smiles and breathes a big sigh of relief. No one is looking forward to another search through the buildings. And it is good to know that they too have escaped.

Rivka is still concerned about all the dead bodies, especially the Seventh Day Baptists who had fought the Gia Union so valiantly. She asks, "Should we bury the Baptists before we move along?"

Yoni says, "Yes, I think that would be appropriate, don't you? I suggest using the craters in the plaza pavement as they are essentially graves already dug. And they will be a fitting memorial."

And so the team drags the bodies into twenty-one craters in the Plaza payment. They cover them with dirt, rock, gravel and loose pavement shards. The whole detail takes about two hours.

Yoni and the entire growing team stand at a distance, raise their hands with palms stretched out, mimicking Yoni's posture as he relates the following short words.

"As the prophet Job said four thousand years ago, 'Oh that my words were now written, Oh, that they were printed in a book! That they were graven with an iron pen, and lead in the rock forever! For I know that my Redeemer lives, and that he shall stand at the latter day upon the earth. And though after my skin worms destroy this body, yet without my flesh shall I see God'" (Job 19:23–26).

Yoni then adds, bellowing out with a deep, echoing voice, which echoes among the buildings, "And so shall it be for these eighty-one brave men!"

All say, "Amen."

"Now, on to Rockford!"

The teams divide the fourteen new crewmen among the three rafts. It is getting tight now. The teams push off from shore as Yoni lifts off, slowly hovering above the water.

Henry, the first rebel, yells to Yoni, "Hey, when will my fallen comrades see God, like you said?"

Yoni smiles and says, "We'll see."

And this time, he lifts off like a rocket. He does a full loop and then a 360-degree spiral, as if he is somehow rejoicing in a victory celebration. Again, he is heading south and west.

The river current is swift.

The rafts are getting waterlogged with over fifteen passengers per raft, plus supplies. They ride low in the water and their movement is more difficult to control. There is more resistance to forward motion and to side steering maneuvers. But the river is straight for the most part and within a few minutes they are beyond Beloit. The view is again farmland.

The Baptist rebels, as they are now being called, are now more talkative. Possibly, for the last few years they have kept a lot of thoughts hidden and became introverted. But now they are loosening up and begin to introduce themselves and communicate with other crew members.

Within two and a half hours the rafts come into the northern suburbs of Rockford. They are now out of Wisconsin and have entered Illinois, although one cannot tell that from the scenery. Rockford looks a lot like Janesville, only bigger.

Burned-out buildings, abandoned vehicles and wrecked boats along the shoreline. No sign of life.

But, no signs of war or battles.

As they pass through, they are on the lookout for another Gia Command Center. But there is none.

The trip through Rockford is essentially uneventful.

Noticing that Isaac seems a little distant, Teah is concerned and sits down beside him.

"Wanna to tell me about the history of Rockford?" she asks.

"Not now." Isaac stares into the distance.

There is a long pause. Teah realizes that if the silence is to be broken, it will have to be her.

"Look, back there at the Command Center, when I warned about saving those people . . . I'm sorry. I may have been wrong. But it did need to be discussed and given consideration. No one had even bothered to notice they were Gia Troopers with The Marck. Even after this long fight with Gia Troopers."

Isaac silently continues to look ahead.

"Want to talk about it?" she asks.

Isaac pauses and continues to avoid eye contact. "Not now, I'm really not upset with you. I just need a little time to process it all."

Teah continues to sit beside him in silence until it is her turn to help with the stern punting chore.

It is about one hundred miles from Rockford to the Mississippi, as the river flows. There are no more major cities along the way. It is essentially former farmland, dairy farms, small towns and resorts.

Joel is thinking about their Rock River experiences and how easy the portaging was for the last three dams they had gone around . . . especially compared to the one south of Lake Koshkonong. Those last dams had very gentle slopes on the edges. Fortunately, in the early 2000s, Illinois made portaging more efficient by establishing standardized systems on all the dams along the Rock River.

Ellis is noticing and commenting on increasing signs of rebounding wildlife. Fish can occasionally be seen splashing out of the water (probably carp) along the edge of the raft's wake. There are turtles and an occasional grouping of starlings nesting in the cliffs. No ducks, no geese, no otters and no beavers are spotted. At least . . . none yet.

But new green shoots on the marsh grasses are now more evident and from time to time a crewmember points to some greening spot in the prairie grasses. The trees are still barren with burned bark and stripped of any overt sign of life.

The crew takes turns at the punting poles and naps in between duties. The rafts have put in a good eleven hours of travel today without any serious delays or events. Although the atmosphere and the water are now clear and there are increasing signs of aquatic and animal life, there are no signs of humans other than those on the rafts. Everyone wonders where have all the people gone?

As the sun begins to set, they all start to look for a good spot to put in. Ian and Miriam take turns screening the river shoreline ahead through Ian's sniper scope. They chat and point to what they are observing in the distance.

Isaac and Matt are tempted with the idea of continuing to drift through the night, since the river is now consistently wide, deep, and generally straight. But Joel and his team, on a smaller raft, are not as confident of this idea. Their raft has more pitch and roll in the current. It is certainly not as stable as Isaac's and for that matter, Matt's. The three rafts agree to pull alongside each other for further discussions of the proposal. Everyone joins in the conversation. Everyone seems to have different reasons why or why not.

Al says, "If Joel's raft has so much side-to-side roll, perhaps we could link the three rafts together. That may provide more stability for all?"

"What's your idea?" Joel asks.

"Well, since some of the river is still too narrow for all three to get through if we are all linked side by side, we could consider another configuration. What if we put the biggest raft, Isaac's, in the front middle, then we link the other two side by side. We could then connect those two to the stern of Isaac's raft. It would kinda be a pyramid arrangement. That layout might provide less pitch, roll and yaw as well, maybe for everyone.

"If it works, we could combine our punting and guiding crews and allow everyone to rest a little more. Hell, we could even rotate crews during the night."

Ian says, "Might work. The moon's now first quarter, so we'll have some moonlight for the first five or six hours. Even if we didn't go all night, we could get in another few hours of travel."

Teah speaks for the women who had already discussed their assessment. "We're in, if there is a general consensus."

After another several minutes of discussion, there are no compelling personal or technical objections offered. So the two smaller rafts now maneuver side-by-side, Matt's to the left, Joel's to the right. Matt then lashes them together. Next Isaac and his team ease the stern of their raft into the common bow of the two joined craft. Aaron ropes the bigger raft securely to the others. Checking the ropes and ties, all seems secure.

The whole contraption looks a bit odd, but all agree that it is definitely more stable for everyone. And it handles very well.

In fact, the V-shape allows it to flow through the water better. There is a reduction in the lateral drag as well and turning requires less effort.

"I suggest we continue as long as there is moonlight," Teah says. "Then we should re-evaluate before midnight and see how we have been doing in the darkness."

Teah's opinion appears to reflect out-loud what the others are already thinking.

With all this activity, no one has commented that the sun has already gone down. There remains only a lingering golden glow on the western horizon. The first quarter moon is very clear, riding high overhead. A bright sparkling white Venus is just above the golden glow from the already set sun. Jupiter is soon visible, a few degrees east of Venus, very bright, but just about one half of Venus' glow. Over on the eastern horizon, on the other side of the sky, is the dusty brownish colored glow of Saturn.

The bright star Regulus, the "heart of Leo," comes into view as the sky darkens. Slowly, thousands of other stars are to be seen. They now shine clearly and brightly. There is no light pollution.

All is quiet, except for the crackling sound of the river's current, and the irregular splashing waves against the bow and sides of the reconfigured raft. Ellis, Al, and Rivka are the first for pilot and punting duty. Teah is lying on her back, looking up at the moon and stars.

Isaac moves aft and sits beside her, then lies back, scooting beside her. He also looks up at the stars.

"What's that star over there?" Teah asks.

This is just what Isaac was hoping she would ask, as he loves astronomy almost as much as history. In fact, he has a thing about astronomy history.

"That's Regulus, the king star. Those two are planets, Venus and Jupiter. Planets have a steady bright glow, and the stars themselves all seem to twinkle. Regulus twinkles, so you know it's a star."

Teah says, "The sky is now so beautiful at night. I really missed the stars during all the pollution. Just look at how bright Regulus sparkles."

Isaac says, "Did you know it's called Regulus because it's said to be associated with the making of new kings? And it lies in the constellation Leo, the lion?"

"Okay, help me get the constellation outline."

"If you look at the triangle of bright stars, that's the body of the lion, and the backwards question mark is his head and front legs. Unlike most constellation names, it doesn't take too much imagination here to vision a lion, does it? See how Regulus is at his heart. Sometimes it's called the heart of the lion."

"Yeah, I can see it now."

"I don't want to bore you but Regulus is a large, young star that's really very interesting.

Teah says, "I would love to hear more. And Isaac, you never bore me."

"I hope not, because I really enjoy our conversations. Well . . . Regulus has three times the mass of our sun. But what is especially interesting is that it spins thirty-five times faster than our sun, spinning at 700,000 miles per hour. If it spun a little faster, it would disintegrate. It spins so fast that its equator bulges making it a broad oval rather than round. Even the light on the equator is dimmed by the much greater gravity at the equator, making the poles five times as bright as the equator.

"And the axis of its spin is in line with its path around the galaxy, making it sort of like a spinning bullet as it travels. Yet, it is so far away, that it has moved very, very little as viewed from earth over the last two thousand years."

Teah gives Isaac a soft look and squeezes his hand. "You may know a lot of stuff, but now you're BS'ing me! How can we know that much about its surface, spin, and shape? It's a long, long way away."

"No, no!" Isaac quickly assures her. "There once was an international program to study the surface of stars. It was called CHARA for Center for High Angular Resolution Astronomy. CHARA combined the images from six large special telescopes spread out on top of Mount Wilson, just above Pasadena, California. The combined image from the six scopes was like having a telescope 1,000 feet across. Regulus was the first star they reported on, way back in the early 2000s. They went on to study similar things in hundreds of other stars, but the Regulus finding still remains the most impressive. It's the first time we could show that massive gravity could affect light."

Teah says, "I don't understand how you can remember all this stuff. You know, the scientific details and all."

"It's easy if you really like the subject, and I love astronomy history. But, what's really interesting tonight is its relation to human history. It's even more exciting than Regulus' angular momentum."

"And, I'll bet you're really right." Teah is smiling, knowing he now wants to talk about history. So she moves the discussion in that direction. "Okay, why do they call Regulus a king maker?"

"Well," Isaac says, "the series of kings whose reigns began with conjunctions of the star would bore you. So I'll skip that, for now. But the list includes kings of Rome, Greece, Babylon, Persia, Akkad, Arabia, and India. That star was called Rex by Ptolemy and named Regulus by Copernicus himself."

"From the times of the Persian Empire, Regulus has been considered to be one of the four royal stars. The Persians considered the four to be associated with the Four Guardians of Heaven. At times in the past these stars have marked the equinoxes and solstices, the beginnings of the four seasons.

"These 'guardians of the heavens' of the Persians, interestingly, happen to be the same characters as the four archangels of our ancient Hebrew traditions. These star associations are, Michael with Aldeberan, Gabriel with Foralhaut, Uriel with Antares, and Raphael with Regulus. And they mark the North, South, East and West. Just think, we've just seen these real Archangels in action!"

Teah says, "I know. History, legends, science, our personal lives, all of it, now seems so intertwined in a gigantic swirl."

"I agree and I don't think we've seen anything yet!

"You know, our tradition tells us that God taught astronomy to Adam and Eve and early men and women. But over time, that knowledge deteriorated into the religions of astrology. These developing new religions, based on astrology, perverted the knowledge of astronomy into systems of secret knowledge, power, control, and oppression. The astrologers eventually controlled the priests, then the kings and leaders, and they in turn, controlled and oppressed the people. Religions, it seems, have always been used as a method of power and control. But I'm getting off track ..."

"Yes, you are. But it's still interesting and it tells me a lot about your views on religion and politics and such. But let's get back to 'what's that star?'"

Isaac grins. "Yeah, thanks. What's really interesting is that this king star, Regulus, and the king planet, Jupiter, crossed each other's path twice in the year 3 BC, near the time of the birth of Christ. Some think that the conjunction was a message to the wise men (from Persia and Babylon) a message that a new king of the Jews had been born. A king star crossing the path of a king planet would suggest the omen of a new king for sure. But why of the Jews? Why a Jewish king? Well the answer is because it was occurring in the constellation Leo the lion, which has long been associated with the "Lion of Judah."

"Okay, you're getting to a big point now, I think?" Teah says.

"Yes I am, but there's even more. Look how close Venus is to Jupiter right now. It was like that in the autumn of 3 BC. In fact, there was a conjunction between Venus and Jupiter, a conjunction that just preceded the Regulus/Jupiter conjunction, later in September. And all that while the sun was in the constellation Virgo."

Teah asks, "Are you saying that you think that you know when Yeshua was born in Bethlehem? 3 BCE?"

"Yes, I think I do. There's compelling evidence he was born on September 11, 3 BCE. Aside from updated and compelling history pointing to that date, the tight correlation with these conjunctions and the star-planet conjunctions we were just discussing, all add up to a fascinating case for that date.

"There's even more. In the Bible, in chapter 12 of the book of Revelation, it tells of the birth of Yeshua in an astronomy-astrology parable. It tells of the coming the new king of Israel to be in the belly of the virgin. Astronomically, that would be Virgo. Now when the sun is 'in' a constellation, you can't see it directly from earth because when we see the sun, it is daylight. What we see is the constellation opposite the one the sun is 'in.' It's only by knowing the positions of the constellations, that can one know where the sun is residing.

"So when Revelation 12 pictures the sun as in Virgo and the moon at her feet, we look to astronomy. The moon crosses our sky, moving from west to east about twelve degrees each night. It would come into the head of Virgo and thru her body during new moon and would appear 'at her feet' as the first visible crescent of the new moon. A rare such thing happened on September 11, 3 BC. I read about that report several years ago. So I immediately went back to my library and ran my astronomy program to confirm it. Yes, the astronomy supposition is true."

"Well," Teah asks, "was that on Rosh Hashanah, by any chance?"

"Teah, you catch on really fast! Rosh Hashanah is also called the Feast of Trumpets in the Bible. That Yeshua would be born on that date makes a lot of sense. Many scholars have predicted such."

"Well, Isaac, did you or anyone happen to notice that just a few days ago, on the day we all saw the hologram of Yeshua's return to Earth, that day was Rosh Hashanah?"

Isaac is startled. For a couple of minutes, he could not speak. "Well, now that you mention it, I think it could've been. It's just that so much was happening. We'd almost lost track of time. Do you think it really was?"

"Yes, it was! It was Rosh Hashanah! We poor little Whitewater students had not lost track of time like you Marines had. It was Rosh Hashanah!" Teah smiles, as she was able to get in that subtle, but amiable dig, a dig at this big-shot Marine.

"Okay, I believe you, but the sky was polluted and no one could see the planets or stars or even the moon. So you're right, I guess we did lose track of time. But to be fair, we did have a lot of stuff, like staying alive, keeping us occupied."

She ignores his defensiveness. "Isaac, if what you are saying is true about the conjunctions relating to Jesus' birth, could the same thing have happened a few days ago, at his return now, as King of kings?"

Isaac had to pause for a while, to think before he could reply. Then he says, "Well, looking at the planets and stars which we can see right now, Venus and Jupiter and Regulus are close to each other. I don't think, however, I can figure it out myself without my astronomy simulation systems.

"Maybe, when we get to Israel, someone will have preserved some astronomy system to help me run a simulation. You'll have to remind me to check it out."

She smiles and looks into his eyes. "I will. It's a date." Isaac is not quite sure how to interrupt her look, but decides to not make any assumptions.

"That's guessing there will be computers, astronomy simulation systems and all that kind of stuff again." They both pause, and have to think about that. Who knows what is ahead in technology.

Teah says softly, "Well, thanks for the astronomy lesson, pal, but we'd better get some sleep. My turn to pilot is coming up next, so goodnight."

"Night," Isaac replies. Then after a half hour of reflecting, he turns over and falls asleep too.

It is now midnight. Aaron, Linda and Miriam have taken their shift as the three helmsmen. The quarter moon has dropped below the horizon. There comes a sudden bump from the right rear of Joel's raft, jolting the entire raft assembly, awakening everyone.

Aaron, the point pilot, yells back to the stern helmsmen. "Dammit! We just bounced off the right bank. Guys, I can't see a freakin' thing. The moon has set and there's no light."

Raquel, just awakened by the bump, yells to Aaron. "I think I hear a distant rumble. Could there be some rapids ahead?"

"I don't know, but in my opinion, it's time to pull over onto the nearest sandbar."

"Agree," comes the answer from the other helmsmen.

Aaron yells, "There's a bar directly to starboard, let's punt to the right."

The raft assembly quickly shifts to the right. It then hits the muddy shore bank with squishy "thump." Everyone is thrown to the right, then forward. The large raft has come to a rest.

"Anyone hurt?" Aaron inquires.

"No, we're okay, comes a voice from the center."

Ellis and Matt jump off into the muddy bank and secure the raft.

Matt yells, "I think there is a sandy bank further ahead, hold on while we walk the raft down there."

Matt and Ellis push the raft back out into the river, loosening it from the muddy bank. The assembly rotates to the right, the left stern shifting back into the current. Joel and Isaac jump into the water and join the effort. All four men now strain to keep a tight hold on the starboard lines. It takes great effort to move it along and keep the stern along the shore. The four of them are gradually able to slide it along until it rests on a gentle slope of a sandy bank.

"Good work men!" Miriam yells.

"Now what?" Linda asks.

Joel says, "Okay, this is obviously all we can do for tonight. But, we've made really good time. I think we've put in at least another twenty to twenty-five miles toward our destination.

"Raquel, I'm not sure, but I think I can hear something too, so we'd better check out possible rapids when we have some daylight tomorrow. For now, let's get some sleep. Anyone who wants to can just sleep on the boats. Ellis and Matt will make a shore camp for the rest."

Ellis and Matt have brought along, on board, a half dozen small limbs and fire tender so they are able to quickly make a campfire on the sandy bank. About half the crew takes their sleeping pads to the sand and set up positions near the fire. Everyone is tired, and all are quickly in a deep sleep. No one even thinks about setting up a night guard with perimeters.

ILLINOIS – DAY 11

The sun's warm morning rays awaken everyone. Isaac is gathering the shore group around the ebbing fire. He pauses and leads in a morning blessing. Then, after a quick snack, they break camp and all are soon ready to get back on board, ready for a day of adventure.

Before pushing off, Ellis reminds the group that they must proceed with great caution and be prepared to act quickly if Raquel is right about rapids.

Ellis has barely finished his warning when Raquel, who seems to have the keenest hearing shouts an alert to the group. "Listen! There's definitely a rumble. I think we've absolutely got rapids ahead."

All are quiet, cupping their ears to better perceive the distant deep rumble.

Al now thinks he can hear the noise from around the distant river bend. "Yeah, no doubt about it."

Michael, Betty and Ian are steering the raft assembly.

"The current's speeding up, for sure," Michael says. "Let's reverse our punting, to reduce our speed and keep good control."

Everyone is alert and intently staring ahead. The raft assembly eases around the sixty-degree right river bend.

"There, there she is!" Linda calls out. "There's the rapids!"

"Let's ease up closer and pull over to shore," Ian says.

All agree. Ellis, Aaron and Isaac pick up the extra poles and assist in easing the raft assembly to the now rocky right shore bank. Matt and Al jump off onto the rocks and wrap the starboard lines around large boulders.

Joel speaks first. "These rapids appear to be the biggest we've seen so far. I know this river and I'm aware of some larger rapids, but I've never been down it this far."

Ian, looking through his sniper scope, concurs. "Doesn't look real good. I see several large, jagged rocks close together and lots of smaller rocks sticking out of the water. The drop off appears to be steep and the current fast. I don't know if we can steer around them."

Ellis says, "I have some experience in rafting the Grand Canyon rapids. It was rough. I remember shouting curse words that I didn't even think I knew. It can really be tricky and dangerous. It's a hell-of-a job to avoid the rocks and if we hit one side of the rocks, it will spin us around and we will lose all control. We could smash up the entire group. I don't think we should ever try to run these rapids like we're arranged."

"So?" Teah asks.

Ellis gives Teah a smile and turns to Al. "What do you think of this idea? Suppose we rearranged the rafts end to end the biggest first, tied to each other about five to ten feet apart? That would allow better maneuvering. And if one of us hits a rock sidewise, the other two rafts would keep the whole group pointing downstream."

Al says, "I like the idea. We could put two poles on either side of each raft, giving six poles to each side. A bit like an insect."

"Well," Miriam asks, "anyone have any other ideas?"

There was silence.

"Then let's do it," Ian says. "I'm anxious to get to the Mississippi today, but in one piece, I hope. All agree?"

Everyone nods or grunts, "Yeah."

Ellis then gathers the group together and shares more of his experience rafting in the Grand Canyon. He explains that the turbulent waters are extremely dangerous and the ride down the river will be very rough and it will be difficult to hold on. Everyone should secure themselves at all times, even when there is an ebb or slowing of the current, changes will be rapid and unexpected. Unless there is an emergency, all crew members should be quiet and must listen to the helmsman who will be shouting orders constantly. Everyone must be alert and ready to immediately respond if necessary.

Ellis then directs Al and Marion to get the store of rope they had scavenged from the ravaged Gia Command Center. The rope is used to attach four lines, running from aft to stern, on each of the three rafts. The two center-lines are to be used by all non-punting crew members, who are to hold on and remain seated close to the center of the raft. The lines running on each side of the rafts are to be used by the pole-punters who are instructed to tie a lifeline around themselves and secure it to the raft's safety line.

Matt and crew first untie their raft from Joel's and both untie their bows from Isaac's raft. Isaac's crew moves downstream about twenty feet while Matt's raft is maneuvered into the middle. Joel's is last. The three rafts are lashed to each other, using three ropes, starboard, mid-ship and port, with a ten-foot space in between.

Isaac, Aaron, Ellis and Ian take poles at each corner of raft one. Matt's and Joel's rafts each set four pole-punters at the corners. All are securely attached to the safety lines.

"Let's go!" Ian yells.

"Go!" Matt follows.

"Away!" Joel affirms.

The three-raft caravan slips into the side current, then moves into the central current and over to the left, where the current is slowest. Within three minutes they are entering the quarter-mile of rapids.

"To the left! To the right! More to the right! More!" Ellis bellows orders to the pole-punters. "Now to the left!"

The bow of Isaac's raft slams onto a submerged rock, which slips under, scraping the entire under surface,

Isaac shouts, "Matt, Joel, push off to the right! Now!"

The two following port-side rafting crewmen push vigorously toward starboard.

"There!" Joel yells, "There's that SOB off port."

They just miss the submerged rock.

"Watch out!" Matt calls, "Big damn rock starboard."

As Matt's middle raft strikes a sharp protruding rock mid-ship, the raft's starboard log bends then breaks, lifting upward. In spite of instructions, there are several stifled yells and groans and a few muttered curses. The seated crew-members frantically hold on to their safety ropes as the impact causes both the front and rear rafts to violently shift upward and to the left.

"Push to port!" Ellis loudly shouts.

The two rafts push off to port and the assembly starts to straighten as there is a brief interval in the violent waters.

"Matt! Anyone hurt? How's your raft," Joel yells.

"Anyone hurt?" Matt surveys his thirteen-member crew. All look around, then at themselves, and nod.

"All okay, just scrapes, bruises and some shattered nerves," Matt responds. "We're still afloat and will have to assess later.

"Good! Just keep pushing."

Another very large rock is just ahead, midstream. It has sharp jagged edges and angry, swirling eddies on two sides and downstream.

"Everyone! Push hard to port!" Ellis yells.

The two rear rafts swing port-side, but Isaac's heavier front raft moves only slightly. Suddenly the whole assembly swings sideways, Joel's to the left, Matt's in the middle, Isaac's now off to the right.

Joel yells, "Dammit . . . we're twisting around backward. Can't stop it."

"The current is pulling us to the right side," Ellis yells. "We're going to hit! Quick! Untie our raft! You all go around to the left. We'll try to go to the right."

Teah, Raquel and Miriam, thinking and acting quickly while there is a little slack in the lines, untie the ropes on the stern.

"Push off!" Ian yells to Aaron and Isaac. They respond and push off from Matt's raft, pushing Matt to the left and giving them momentum to the right.

"It's gonna be close!" Ellis yells. "Guys, before we hit, push off the rock as hard as you can."

Joel's raft has now swung around backward, pulling Matt's backward, but missing the large jagged rock off to the left.

"Ooo rah," the four Marines start to yell, as they push against the rock. As Isaac shoves with all his might, his poll skates across the slippery algae coated rock. Losing his balance, he hits the water just as the raft swings around to the side.

Aaron screams, "Man overboard!"

A cold chill grips everyone as they vainly search the surface looking for Isaac, who has been sucked to the bottom by the chaotic but coordinated downward pull of the swift current sweeping around the rock blockade.

Isaac is now a prisoner of the furious whirlpool. He holds his breath and violently thrashes upward. He thinks, "Need to swim downstream, away from this whirlpool." He twists his body to face downstream, rather than up, and desperately holds his breath. Just as he can hold his breath no longer the current begins to ebb. On board all are searching the surface, and pushing their poles into the whirlpool. Thirty seconds go by, forty, one minute. Then, twenty feet downstream Isaac's head pops through the surface. Then back under again. The rafters shove the raft forward as hard as possible.

Just before lunging for the surface again, Isaac feels a forceful pull around his waist. Almost immediately he is hoisted aboard by Ian and Teah, the first one to grab Isaac's safety line. Everyone kneels over Isaac, placed face down on the raft, as he coughs up inhaled water, then begins to breath smoothly. They roll him side to side to assist bronchial drainage. Finally, he sits up and smiles as he regains his composure and notices forty-nine apprehensive but grateful and relieved companions staring at him. He is soaked and more than a little embarrassed.

Isaac says, "I'm okay! Thanks for pulling me aboard! Playing like a sea-anchor ain't much fun! Imagine going through years of fireballs, earthquakes, plagues of all kinds, then drowning in a little whirlpool!"

Then to break the tension, Isaac screws up his face and calmly asks his team an obvious question.

"Has anyone noticed we're going ass-backwards?"

Several laugh and Ellis shrugs his shoulders. "But it's a rectangle, shouldn't make any difference."

"You guys okay?" Teah yells to Joel and Matt and crews.

Joel calls, "Yeah! But let's pull over below the rapids to reassess."

"Hallelujah!" Al cries, "hallelujah, hallelujah."

"Amen," is the multiple response.

Just beyond the bottom of the rapids the river widens and there are sandy backs on both sides. The three rafts push ashore on the left bank, side by side.

"Everyone off," Isaac says. "Let's check the rafts and each other."

After ten minutes of survey of the rafts, they find Isaac's and Joel's are without substantial damage.

"Matt, what do you think of your right-side beam damage?" Isaac asks.

"The beam is broken, but the raft appears generally stable. It'll tilt to the right, I think. I suggest we lighten it, if you two could take our supplies and maybe a passenger or two."

Al says, "If we now go back to our original configuration, Matt's damaged right side will be supported by the left side of Joel's."

"Yeah, that makes sense," Joel says. Isaac and the others concur.

Raquel reports on the status of the crews. "No one, not even Isaac, seems to be seriously hurt, just shaken nerves and a few scrapes."

The crews maneuver the three rafts back into the triangular configuration. Two members of Matt's crew are moved to Isaac's raft and the supplies are redistributed. The triangular raft assembly is pushed back into the current of the Rock River.

"The way I've estimated this," Ellis says, "we're only about twenty miles from the Mississippi. Providing there are no more unexpected adventures, we should be there by early afternoon."

Although no one really knew what he meant by 'unexpected adventures', still they yell, "Hurray!" And up goes a general cheer, "Ooo rah," from the four Marines.

"Hallelu, hallelu, hallelu, hallelujah, praise ye the Lord," sing the fourteen Baptists. Al, Betty, and Linda join them in the old hymn of praise.

The river is flowing faster as it gathers more and more water from its tributaries along the way. But no more rapids. And no more dams. It is a time of anticipation.

Just after noon, Ian spots the high cliffs of the great Mississippi Valley on the horizon. Everyone's pulse rate is increasing. Their decision is to not stop for lunch or for anything else. The remaining berries, roots, bread and vegetables, which Yoni had left, are spread out for snacks as needed. Anyone who needs to pee will just have to do it overboard. The rafters are now obsessed with getting to the Mississippi, whatever that will mean to them.

The high cliffs of the Mississippi Valley began to dominate the horizon. Soon they fill the right and left views as the team approaches the mouth of their now personalized Rock River.

Ahead is open water, the open water of the mighty Mississippi. Compared to the rivers they had been conquering, the Mississippi is in a league of its own. Here, it is over a half-mile wide, with towering cliffs on either side, and a marsh island on the western side. The spectacular bridges of I-280 span the river a half-mile downstream. The view itself is overwhelming.

But then, the surprise!

CHAPTER 11

BIG BOATS

Ellis, often the first to speak, shouts with his utmost volume. "BIG BOATS! BIG BOATS!"

Yes, everyone can see them. Anchored about a hundred yards off shore are five giant white glistening ships, each over six-hundred feet long with giant silver reflecting windows and streamlined bodies. Each is capped with a giant glass-windowed bridge and a forward-windowed deck.

Creating a dramatic contrast, along the Mississippi shoreline are more than a dozen rafts, each with its own custom character. Hundreds of people are milling around on the shore watching six tender boats shuttling people to the ships.

Isaac, Joel and Matt take control of the raft piloting. They aim to disembark just downstream from the other rafts on the eastern Mississippi shore.

Raquel yells, "Look! Yoni! On the shore!"

Indeed, it is Yoni. With him are two others, who look a lot like him. One appears male, like Yoni, the other appears to be a female version of Yoni. He is discussing logistic details with his companions and several others.

Raquel waves at him. He looks up and waves back. He motions to come over to where he is.

The team docks just downstream from Yoni. He breaks off his conversation and, along with his companions, walks over to give a hand to each member of the excited crew coming off the combo-raft.

He greets each one by name and gives everyone a big hug. When all are off the raft and on the shore, he introduces the two other Kadoshim. "Chavah, Natan."

Yoni walks over to Ellis, puts his arm around his shoulder. "Ellis, are these rafts big enough?" With his characteristic smile and a wink, he points toward the Big Boats. Everyone smiles and nods. The joke and truth is on Ellis.

Then comes the process of embarkment, sorting out who is going to which boat and determining what supplies are needed. Yoni, Natan and Chavah seem to have everything well organized.

Natan informs the group. "Your raft teams will be kept intact. You should choose a designation, a name or an identity for your group, to help with our coordination once you are on board."

The team members were already trying to decide on whether they are the Bark River group, the Rock River group or the Wisconsin group. No decision is needed at the moment.

Teah sighs and takes a deep breath, followed by a spontaneous and relaxed smile. This catches Isaac's eye. He understands her thoughts and her emotion. They will continue to travel together. He too breaks out into a big sheepish, embarrassed and juvenile grin. They both now feel that this is destiny. They feel they are together for a greater purpose.

Chavah continues the instructional orientation. "You do not need to bring any personal effects on board, unless you have something you just can't part with. If so, bring it. We are not that space limited. But everything is provided, food, supplies, and reading materials. Each of you will find seven units of clothing, custom-made for your preferences. If anything does not fit, let us know and we can make alterations. If you do not like our choices or styles, same comment.

"Now, time is passing quickly and we really do have a tight schedule. We would like to get underway before sunset. We hope you will enjoy your next sunset from your friendly Big Ship!"

Ellis asks, "What about the rafts. We've sort of become attached to them."

Yoni, appreciates Ellis' dry humor and thinking. "We are leaving a great deal behind. A lot of people once inhabited this great land. We are leaving their remains, the great cities, the farms, the small towns, the expansive frontier, the rivers, the mountains and the great monuments. All this, we are leaving behind.

"Perhaps someday, some young archeologist will find the buried remains of these wonderfully handcrafted custom rafts, buried in the sand or sandstone near the shore of this great river. Who knows . . . he may write his master's thesis with his own speculations.

"He may speculate that two civilizations clashed in the twenty-first century. Or that technological civilization had not reached Wisconsin at that time. Maybe he will speculate that the Packers were the origins of Wisconsin civilization, since the big green and gold G is everywhere. Perhaps he may find all these rafts and write in a journal how he has found the raft-building center for this civilization. Too bad we do not have an old iPad to leave on board. That would really be interesting. We could go on forever on this.

"Ellis, did you ever hear of the old, long out of print, spoof book, *Motel of the Mysteries*?

"Yes, I have. Grandpa had a faded copy in his library. I remember it."

Yoni says, "I thought so. Well, that says it all.

"Let's not play with this future archeologist's mind. But let's do leave him this material." Yoni cannot resist an uncontrollable smile, even if smiling at his own witty remarks.

Yoni continues to smile as he looks directly at the four Marines. "But I suggest you bring your journal to read to your grandchildren, rather than leaving it for the future archeologist."

Yoni then addresses the whole group. "This is an emotional moment for us all, and it is important that at this time we should pause, reflect and remember.

"All is okay. In Hebrew we have a saying, *'HaKol b'seder.'* It means, 'The all, the everything, the all is in order.' Today we just say, okay.

"So today, my friends, *'HaKol b'seder.'* Any questions?

"Any questions, Ellis?" Yoni has become close to Ellis' somewhat skeptical, but honest way of thinking which is not too unlike his own personality. Ellis slowly shakes his head.

The tender boats are now here for the newly arrived fifty Rock River passengers. The team is to join several hundred others on board the *Millennium Eagle*. As they cross the three-hundred-feet of water, there is total silence. Everyone is in deep thought. Some are quietly praying. No one can speak. Several of the team members look back at the shore, focusing on the three rafts tethered together, just downstream from the others.

Finally, Miriam says, "Thank you, you piles of wood and vine."

Linda also breaks the silence. "Look, there are big birds soaring overhead and they are not buzzards. They are eagles!"

Aboard the *Millennium Eagle* there is a buzz of activity.

Yoni comes on deck to speak to the arriving passengers. "You are joining several hundred brothers and sisters who are already on board. Some embarked today, here at the Rock River. Others embarked over the last two days. They have come from up-river, from the Dakotas, Minnesota, Michigan, northern Wisconsin and some from Iowa. Of these hundreds, some have been assigned to the *Millennium Eagle* and others assigned to one of the other vessels. Assignment is not just a random number. There are important and individual purposes involved.

"As you will soon note, this vessel is now only twenty percent full. As we progress down the Mississippi, we will gather others at the junctions of the Missouri, the Ohio and the Tennessee. And we will get additional passengers at the mouth of the Arkansas and the Red Rivers and from the many other tributaries of this great river.

"When we get to the Gulf, even more will join us. Along the journey, we will be joined by thousands of others in ships from North and South America and the Caribbean. It will be easy to be overwhelmed by the thousands of Sephardi from the Caribbean with their unique Hispanic cultures. Ultimately, we will be joined by other vessels from Africa and Europe.

"Welcome to the voyage home. The next few days are going to be really exciting. They will be splendid. As our Jamaican brothers say, 'Don't worry, be happy.' The Kadoshim know the details of how many and who will be rescued at each point. There is no uncertainty. Yeshua, from Jerusalem, is in full control. He knows every step, every person. Not one of his little ones will be left. Not one of his sheep will be lost.

"As Yeshua commands us, 'Comfort you, comfort you my people.' That is our command, our mission. That is our goal. That we will accomplish."

Throughout the ship there is now a long silence, as every passenger reflects on the profound message of these statements.

Isaac leans back and awkwardly takes Teah's hand. "As he says, each of us is special. No one can take us out of his hand."

Teah smiles at him and nods, but she is not quite sure what he means.

Chavah comes on deck and is next to address everyone. "I'm sure everyone is wondering . . . where did these big boats come from? They are so far removed from the hand-made rafts you started out with.

"These big boats were designed to be Gia Union Troop Cruisers, providing rapid deployment for Gia Special Ops Troopers. They are the best, the fastest ships ever built. But remember they are designed to rapidly transport troops, not for a luxury cruise of the Mediterranean. No swimming pools, no casinos.

"Now please don't be afraid of them, we have cleansed and blessed them. They are not unclean. They will serve us well.

"If you are just coming aboard, welcome. Now you may be surprised to notice that the operating crew is minimal. We have four professional naval officers operating the ship.

They are technical and navigation specialists. What is unique, for you who have been on cruise ships before, is that every passenger here is also crew. This is not too different from your rafting experiences.

"We are passengers and we are crew. Each of us will have responsibilities for the operation of the ship and be of service to other passengers. So we will refer to everyone as crew. We are all part of this process. Each of us is an important part. As you will see, this theme is a bonding principle of the Kingdom of God, which has now come to Earth.

"I want to share some reflections from our ancient prophets, helping us to keep everything in focus. These ancient prophecies are being fulfilled here, and around the world, today.

"The prophets of Yehovah have proclaimed this day. Isaiah and Zechariah and Jeremiah and Micah and Amos and others all looked forward to this day. From a time when they were seeing the destruction of Judah and of Israel and the temple, they paused, and then looked forward to this day, our day.

"Here, I will read from the Bible that was given to Isaac. Isaiah wrote in chapter 60,

Arise, shine; for your light is come, the glory of Yehovah is risen upon you. Darkness shall cover the earth, and gross darkness the people: but Yehovah shall arise upon you, and his glory shall be seen upon you.

And the nations shall come to your light, and kings to the brightness of your rising. Lift up your eyes, look about, and see: They gather themselves together, they come to you. Your sons shall come from far, and your daughters shall be nursed at your side.

Then you shall see, and flow together, and your heart shall marvel, and be excited; because the abundance of the sea

shall be assembled unto you, the might of the nations shall come to you.

A multitude of transports shall carry you, the transportation systems of Midian and of Ephah. From Sheba shall they come. And also, they shall bring gold and incense; and they shall show forth the praises of Yehovah.

Surely the coasts shall wait for me, and the ships of Tarshish come first, to bring your sons from far, and they bring their silver and their gold with them, unto Yehovah your God, and to the Holy One of Israel, He who glorifies you.

"And there is more. Many prophets looked toward this day. And then from Zechariah, chapter 8, I read,

Thus says Yehovah of hosts; 'Behold, I will save my people from the east country, and from the west country. And I will bring them, and they shall dwell in the midst of Jerusalem: and they shall be my people, and I will be their God, in truth and in righteousness.'"

"Now here, in Jeremiah, chapter 31,

'Behold, I will bring them from the North Country, and gather them from the coasts of the earth . . . a great company shall return. They shall come with weeping, and with supplications will I lead them: I will cause them to walk by the rivers of waters in a righteous way, in which they shall not stumble. I am a father to Israel, and Ephraim is my firstborn.

Hear the word of Yehovah, O you nations, and declare it in the coastlines afar off, and say, He that scattered Israel will gather him, and keep him, as a shepherd does his flock.'"

"Crew members! What you are a part of today, my brothers and sisters, is the fulfillment of what those prophets saw. They saw through a glass darkly. Now today, we see it all clearly. We not only see the prophecies being fulfilled, we are part of the fulfillment.

"Others, people who have survived the tribulations of the last days of mankind's rule on earth, are also being cared for as you are being helped. They will join us all in building the Kingdom of God on earth.

"Today they know what God had promised to Abraham and his descendants, 'I will bless them that bless you, and curse them that curse you.'"

Isaac could not resist asking at least one question. "Then there are others, other than us, of the remnant of Israel, who have survived the tribulation, the plagues and wars?"

"Yes," Natan replies, *"the people of the nations? Yes, lots of them."*

Isaac asks, "Where are they?"

Yoni smiles at Isaac. "We'll see."

Isaac figures the time for more probing questions is not now. For now, he will just have to live with this shortened version.

Ellis whispers to Isaac. "That's good news, as long as they're not Gia Troopers."

Yoni looks at Ellis and Isaac and shakes his head. "They are not Gia Troopers, at least not many."

Now everyone sets about to find his or her quarters. They are adequate, just adequate. Space is ample, but none is wasted. All are supplied with sets of new garments, all custom fit. The garments seem to reflect the personal and preferred styles of each person. No uniforms.

There are bathrooms and shower facilities in each hall. Exercise facilities are available on the main deck. But there are no computers, no televisions or telephones. Those are not possible, for now, as most satellite communications are still out.

The bridge only has radio and video communication with links to each ship and the coordinating centers in Miami, Gibraltar and Haifa.

Duty assignments are found in each person's room. Not surprisingly, the duties fit the skills and interests of each person.

Just before sundown the ship's horns sound. The engines start. The anchors are hoisted and the Big Boats begin to move downstream.

All the crew members move to the large thirty-foot windows to watch the process of getting underway. A big cheer goes up as forward movement of the Big Boat is perceived.

Yoni is now aboard the *Millennium Wolf,* and Chavah remains aboard the *Millennium Eagle.* She comes on deck, blows a shofar and announces, "Dinner is served."

Everyone assembles in the large dining area on the top deck. There are great views forward, as well as to port and starboard. The dining room is, however, only twenty-percent full. There are lots of empty tables and chairs.

The main meal consists of vegetables, salad, grilled perch, coffee, water, bread, cheese, fresh-chilled watermelon and red wine.

Aaron, Al, Linda, Rivka, and Miriam are on duty as servers.

Rivka is appointed to lead the Ha-Motzi blessing, beginning the meal. Several of the Rock River crew, as they are now designated, join and assist her. After dinner, a prayer of thanksgiving is given by Henry.

Then the crew retires in small groups to the forward lounge. There, more wine, coffee, juices, fruit and cheeses are available. There is lots of discussion about the other people of the nations, which Chavah, Yoni, and Natan have just enigmatically mentioned.

"Where are they?" Ellis asks those around him. "We have not seen any, except for the small Greek crew aboard this vessel. I suppose the crews of the other vessels have non-Israelite crews too. But where are all the others? In our travels through city after city, they were empty."

Aaron says, "Being non-Jewish ourselves, Ian and I are especially eager to get some answers . . . but I guess we'll just have to wait and see, as the man said."

While the men are discussing such things, a group of about fifty women gather around Chavah and engage her in conversation.

Teah, taking the lead, looks at Chavah. "Your name is Chavah. That's Hebrew?"

"That's me," Chavah replies.

Teah asks, "And Chavah is the Hebrew name that is translated in English as Eve?"

"Yes."

"Eve, or in Hebrew, Chavah, was the wife of Adam, and the mother of all living."

"That's me."

"Wow!" Teah exclaims.

Raquel says, "I would think you would be in Israel, sitting on some throne or something, with hordes of maid servants and lavish quarters?"

"No. All of the elect, those in first resurrection, are raised to serve, not to be served, not to lord it over others," Chavah says.

"Do you remember all the things in your life?" Betty asks.

"Yes, all nine hundred years of it. It wasn't all fun and games. Actually, if you know your history, I sort of messed up some things.

"Life was tough, especially when Adam and I were expelled from Gan Pardes (the Garden of Eden). It took a while to get reorganized. We had two sons at first. Then one son killed the other.

"Life then was exciting, but full of good and bad. I guess we had it coming. The name of the tree was really quite prophetic.

"Adam was a good husband, but he needed a lot of emotional support. He had a lot of stress too. And we had a lot of kids, grandkids, great grandkids, and so on. By the time of my death, the earth was filled with over ten million of our descendants. We really did know how to 'be fruitful and multiply.'"

Teah says, "Well, it is good to know you, to whom we all owe so much. It is good to know you are among those elected to be in the first, the better resurrection. But one thing is still puzzling me."

"What's that?"

"I thought that those resurrected as Kadoshim would be sort of neuter. I mean, would not have a gender, I mean, they are not men or women, I mean . . ."

Chavah interrupts her. "That's a general perception, not just yours. It seems to be based on a misconception of a reply of Yeshua to a trick question from some Sadducees. They are attempting to trick him about a hypothetical woman with seven successive husbands, brothers who each died in succession.

Let me explain.

The Sadducees, who do not believe in a resurrection, are attempting to trick him by showing how stupid the idea of resurrection is. How hard it is to sort out all of the human relationships. The Sadducees have loved to put this trick question to their Pharisee competitors. And it stumped them every time. It would certainly be a Sadducee polemic victory over Yeshua. They are confident.

Yeshua's response surprises them. It is that in the resurrection, there will be no marrying, nor giving in marriage. Now, everyone jumps on this with a conclusion that there is no gender, male or female in the resurrection.

Yeshua is not saying that. He is not saying that our resurrected bodies will be neuter. But what he says is that we will not be seeking mates, arranging marriages, marrying, divorcing, remarrying, exchanging mates on demand, and so forth. That means in the resurrection, we perfectly formed glorified beings will not be running around in the dating, searching and match-making schemes like before. Competition has been a major focus throughout earth's history.

Finding a mate, competing for a better mate, the ever trying to make that perfect match, a survival of the fittest, all these have been driving forces in human history. Instead, all the matching will be from an all knowing, all good God. There will be no mistakes. There will never be a need to look for a better mate. There will be no need to ask for someone's hand in marriage. No need for a perpetual looking for someone better, a better fit or more excitement in someone else. There will be no such 'marrying and giving in marriage' as a part of life in the resurrection.

The mate assigned by the Father is perfect; a perfect fit for eternity. We are to forever grow in our love and affection for each other. The Sadducees' ignorance of life in the resurrection is exposed. God is in charge of everything. The resurrected do not need to seek or compete for or to figure out a best mate. God is in charge.

Now the truth is that Adam and I are really a perfect fit. As you might say, it is 'a match made in heaven.' And it still is.

Look into your Bibles. Look at the matches made by God, Adam and Chavah, Abraham and Sarah, Isaac and Rebecca, Zechariah, and Hannah, Aquila and Pricilla. They had such good results. Contrast that with those made by human scheming and manipulation. Look at the mess Jacob made and at the results of the constant marrying to build power and prestige like David, Solomon and the kings. Their whole business was centered on power and control through marrying.

No, God is not against marriage, on either side of the resurrection. He is against a life of competition and struggling to better oneself, of stealing another's mate, or just using her or him. Don't we have a commandment about that?

Think about this, our new Kadoshim bodies are glorified. We have spiritual bodies. We are not dependent upon air or water or food for sustaining life. In fact, our bodies are independent of the material world in sustaining our life. Each of us is directly nourished by the spirit of God flowing in and through us. It is our air, or in Hebrew, the 'ruach.'

We are sustained by his *ruach*, just as humans are sustained by the physical air, which is by the way, also known as ruach. We are all linked to Him.

Teah says, "But, you just had a meal with us. You just ate fish and vegetables and drank wine. I am not getting it."

Chavah answers, "We are not dependent on water or food or air. But that does not mean we cannot be a part of the physical world. We can eat, drink, take a deep breath, and talk with real words, vocalized via real vibrating vocal cords. Our spiritual, glorified bodies are not dependent on material things, but we are not limited from them.

"We can live in this material world and fully interact with it. We can even transform it. They call that a miracle. We can also fully enjoy the good physical things God has created. The creation of a male and a female gender is, incidentally, part of that good.

"Why would God, at the peak of his creation, suddenly eliminate a part of it which he calls very good?

"I think this will take some time to digest and we have plenty of time. I will try not to overwhelm you too soon. We have so much to discuss."

"Just one more question," Betty asks, "where's Adam?"

"He is assigned to help in southeast Asia right now, but don't worry, we are not limited by time or space either. I will see him tonight and will give him your regards."

Chavah smiles, winks, turns and then disappears through the forward doorway. The women lean back, take another glass of wine and, for once, are speechless for minutes. Then they begin to rehash it all again.

Miriam says, "Well, that's all so fascinating. We've so much to learn. This is so exciting."

Everyone is now really tired, but very enthused. So many continue to talk and visit until well after midnight. There is no curfew ... only a duty schedule.

DAY 12

Morning begins with music from speakers located in each room. Everyone is thrilled to get up, to work and to learn more. There is an unrestrained spirit of adventure on board.

After breakfast, in the common dining room, Chavah is back and begins her announcements.

"We have made good progress. I guess you have noticed that these Big Boats are hydrofoils. They are cruising at close to top speed. We are now approaching Hannibal, Missouri, where we will be dropping anchor to pick up more of the remnant of Israel. We'll be in anchorage for only an hour, so don't plan on any shore leave, unless you have assigned duties on a tender boat.

"Now it's time to start tours of the boat. All tours will be in groups of thirty and will be scheduled throughout the day. Tours will begin on the bridge so that you can meet the captain and navigator."

Isaac and Teah are fortunate to be in the first group, right after breakfast.

Captain Michaelpoupolos greets the group and begins his welcome.

Welcome to the bridge of the *Millennium Eagle*. She was built by the Helena Company's subsidiary in Davenport, Iowa, in a special government contract with the Gia Union beginning seven years ago. A big order of another eight-hundred of such ships were commissioned, but only five hundred were built in different shipyards around the globe. Each cost 800 million Gia Units.

Two hundred vessels had been delivered to the Gia Union when they then canceled the project last year after the first fireball hit the Mediterranean. We found two-hundred-fifty of them still in various dry docks. Many of them were damaged. Only eighty of them are currently now operational, many of them from the Davenport and Mississippi shipyards. Man, we've been busting our butts getting these cleaned, operational and in place for you all today. Not that I'm complaining. Never thought I'd get to captain one of these, since I, like you, have been on the run for quite some time. I hear there are others units in Panama, Denmark, Australia, and Japan. More are coming online in every day.

All vessels have a basic and universal hull design, but there are individual differences and special designs to fit special missions. They all have the latest technologies and they are all hydro-foils capable of cruising at forty knots. They are not the biggest ships ever built, but I say that they are just the best. They were built as Gia Trooper transports for rapid deployment, there are no staterooms and no private toilets or showers. As you already know, you have to go down the hall for that. But there is lots of common space and great views. Each vessel can carry up to one thousand people.

They are not cruise ships, but they are fast and have an unlimited range. The power units are cold fusion systems. These powerful units run everything, but there are backup batteries and fuel cells. Since the cold fusion source is hydrogen, which we extract from water, our supply of fuel is limitless.

They were not originally called the Millennium series, to state the obvious. The Gia Union called them Gia Cruisers. But now the name of each ship is a Millennium something. As you can see, there are five in our current group, The *Millennium Eagle*, the *Millennium Wolf*, the *Millennium Ephraim*, the *Millennium Ram*, and the *Millennium Dan*. Other big boats will join us when we get to the Gulf and the Atlantic.

Not all our systems are operational, especially, the communication systems. We have had to go back to old ham radio technology for ship-to-ship communication. Although some video transmission is possible, that is limited to the bridge for now. After all the tours are over, you are welcome to return to the bridge if you like. The view is spectacular and we enjoy the company.

Teah just has to ask, "Captain, are you Hebrew also?"

He pauses, and thinks a minute. "Well, if you understand the math of Abraham, everyone on earth has some of Abraham's DNA. Some have more than others. Indeed, through Abraham, Isaac and Jacob, the whole world has been blessed. Now how that blessing translates to our relationship with Israel is a long discussion, one that I am learning myself. Please come back, it is one of my favorite topics and I would enjoy more dialogs on this and other topics. But it is time for the next group now.

"Thanks everybody! It is an honor to have you all on board!"

The tour moves on to the engine room, the turbines, and the cold fusion units. But Teah cannot get her question out of her mind.

Isaac says, "I won't let you forget to ask it again, Teah." He smiles and they go on to the next station.

One hundred twelve more travelers are picked up at Hannibal. Most of them join the *Millennium Dan*, but fifteen are assigned to the *Millennium Eagle*. Two of them were of special interest to the Rock River team.

A voice shouts across the front deck. "Henry? Is that you, Henry?"

Henry looks up, but cannot believe his eyes and runs over to the newcomers.

"Tom! Bart! Can this really be you? Oh thank you sweet Jesus."

The three join in a long, tight group hug, patting each other on the head and shoulders.

Natan comes over. "I guess I do not need to introduce you two?"

"No, no need," Bart says, "but thank you, thank you, thank you!"

The two missing Seventh-Day Baptists have been found!

Henry asks, "How in the world did you two get here?"

By this time all fourteen of the Baptist members and the rest of the Rock River team have assembled to watch the reunion and hear Bart and Tom's story.

Bart was the first to start. "We started as snipers on top of building number eighteen. During the night, someone took out an F35 heading straight for us. It crashed just in front of the building. We thought for sure it would explode at any second, so we decided to take the risk and move out into the open. We were defenseless and there was nothing more we could do without cover. So while God was still giving us the protection of darkness, we decided it was time to head to the woods."

Tom says, "We slipped out of the plaza into the woodlands of Roosevelt Park, then we started toward Milton to check on our families and get them into to hiding. You know, we had just blown up the Gia Center and we figured there could be a fierce retaliation by the Troopers.

Bart turns a little pale and looks down at the floor. "We got there at sunrise. Oh. Dear Lord! There was no one to be found in town. All were missing."

Tom says, "We didn't know what to do or where to go. So, we started back toward the Gia Center and laid low in the woods for a while. By then, when all was quiet and we thought the troopers had moved out to Har Megiddo, we moved in closer. There were bodies everywhere."

Aaron asks, "Did you check for survivors?"

Bart continues to stare at the floor. "Yes, the best we could. There was just the two of us. It was really spooky and the devastation was so bad, we figured that we were the only survivors of our group."

Tom says, "As we searched around, we luckily found an abandoned personnel carrier that was operational. It started. Then we headed south on I-39, not sure where we were eventually going. We ran out of gas, just north of Normal, Illinois."

Bart gives a big sigh and looks up directly at Natan. "That was where this guy, this angel, shows up."

Natan says, "Thanks for the promotion. However, I'm very satisfied to be your Kadosh."

Bart's color returns as his regains his composure. "Somehow, Natan had gas for us and told us to keep on I-39 south to Springfield, then to take I-72 west to Hannibal. We were to meet with others there and some big ships—"

Tom interrupts his friend. "To be rescued and given a way outta that mess was wonderful . . . but when Natan actually removed The Marck from us, it was a miracle! We had been saved!

"Hallelujah," Bart says.

And the reunion of the Seventh-Day Baptists, formerly Gia Trooper commandos, continues all day and well into the night.

Yoni is listening closely to all of this too. He moves up to stand beside Teah. "There are your other two Baptists. You can update your list now."

She smiles, turns and gives Yoni a big hug.

DAY 13

The next day, two more boats join the convoy at St. Louis. In addition, sixty more people, in small groups, are added to each boat. The dining hall is now beginning to fill.

At Cairo, where the Ohio and Tennessee Rivers empty into the Mississippi, three more Millennium vessels add to the fleet, along with four hundred more people, who are distributed to the existing boats.

DAY 14

More are added in Memphis and at the mouth of the Arkansas River. More join at Vicksburg.

DAY 15

By the time the ships arrive at New Orleans, the convoy has grown to sixteen ships. Eight hundred additional people come on board the ships in New Orleans.

By the time the fleet leaves the Mississippi and moves into the Gulf, they are nearing full capacity with 14,635 people on board.

DAY 16

The next stop is Tampa Bay on the southwest coast of Florida. Embarkation of passengers there is extremely difficult since the entrance to the bay is almost completely blocked by the tangled cables and remains of the majestic Sunshine Skyway Bridge, which had previously spanned the open waterway. It is obvious that the area has suffered some massive destructive force. In spite of the obstacles, two hundred seventy-eight more survivors join the boats. Life on the boats is now extremely busy. Everyone has a full list of duties, although they are shared equally.

The mixture of passengers from different cultures is adding to Teah's responsibilities, as the *Millennium Eagle* duty list coordinator. The fact that she has a natural gift for organization has not been lost on the Kadoshim. They respond by giving her opportunities to challenge and sharpen her skills, developing them for even more exciting responsibilities. She meets daily with Chavah and they are becoming good friends.

All three Kadoshim seem to move freely among the ships, advising when needed.

There are occasional accidents, falls, strains, and frequent episodes of sea sickness. Chavah has set up an infirmary in the former ship surgeon's quarters. Raquel has been asked to help there. Volunteering enthusiastically, she is put in charge of the triage. A physician from Puerto Rico has joined the crew and is there to help Raquel as needed, but he has little serious business.

Ian has developed a friendship with the ship's Navigator and is given duties on the bridge, even occasionally taking over some shifts of navigating.

Isaac is spending much of his free time in the library. He has decided to take the opportunity of reading the Bible again, this time the whole thing.

It is strange for a troop transport ship, but there are original language editions, lexicons and lots of history volumes in the library. However, there are no computers or electronic search systems. Isaac's perspective is now quite a bit different. He figures it is also time to brush up on some Hebrew syntax and vocabulary.

Ellis has been assigned by Yoni as the fleet communications coordinator. At first, with only five vessels and five hundred people, it was difficult enough, but the job of keeping communications up to speed is rapidly expanding as the fleet and the passenger list grow. Aaron requested to be assigned to help. Later, Rivka and Miriam are added to the communications team. Since both speak Spanish well, it greatly helps with the additional passengers from Mexico, the Caribbean, and South America who have joined the fleet.

Joel and Matt have engineering degrees and ask to help in the cold fusion and power systems. They are joined later on by several other engineers with experience in nuclear and fusion plants.

Al, Linda, and Betty request duties in the nutrition and food service areas. Everyone's job grows as the population of the fleet swells.

Sailing from Tampa Bay to Miami takes only one day. The west coast of Florida is shallow water and the scenery is quite bland, with beaches, islands, mango marches, and river inlets. As Ellis is strolling along the upper deck of the *Millennium Eagle*, he recognizes a familiar voice. He shoves his way to see the face.

"Shulla? Shulla? Shulla Coplan?" He blurts out.

"Ellis! It is you, Ellis?"

The two cousins are overjoyed to find a family member. They engage on a prolonged, gratifying embrace. When each has his and her breath, they step back to size each other up before being able to speak.

"Did you come on board with the Tampa Bay group?" Ellis asks.

"Yes, am I glad to be here. Don't know how I survived these last three years. How about you? I thought you were in the Marines. That your platoon was wiped out. Wow! Guess that's not so. Where did you get picked up?" Shulla continues to ask question after question.

Finally, Ellis takes her by the hand and says, "We have so much to talk about, but I'm on duty as the communication officer. Let's check by my office and get my staff to cover for a while. If you want, we could take a tour of the Boat while we catch up."

"Oh yes, please, you lead the way!"

They head upstairs to his little cell-like office. Rivka is there. "Rivka, Shulla. Can you believe it? I've found my cousin Shulla. She just came on board at Tampa Bay."

It just occurs to Ellis, where is Shulla's husband, Dan? He pauses and asks the question, expecting the worst.

"Well, I think he's gone, but it's a long story. Let's take that walk and I'll fill you in."

"Ellis, I'll cover for you," Rivka calls out.

Shulla, like Ellis, was also from eastern Wisconsin, growing up on the north side of Milwaukee. She is twelve years older than Ellis. After graduating from Northwestern, she moved back to Milwaukee for a career in finance. She met Dan during a business negotiation. The two were well suited for each other since they both loved to debate and challenge. They married four years later, but chose a career pathway rather than family. They were happy and traveled widely and often.

They find a quiet place to talk and Shulla begins to update Ellis.

Eight years ago, Dan and I had grown very tired of the cold, unpredictable winters back home. So we contacted some of our business friends in Florida. It wasn't long before we both accepted new positions in the high-rise business district of Miami. It was much warmer there and we loved it. However, after we moved, we sort of lost contact with the family back in Wisconsin. On occasion, I did hear bits and pieces of some of your Marine adventures, as much as I guess they let family know. Dan and I were very absorbed in our new business. We were at the center of American commodities futures and it seemed the sky was the limit to our success.

When the United States was absorbed into the Gia Union, Dan and I were at our peak. We arranged vast contracts for American resources on the world market. We became very wealthy, more that we could have ever imaged.

Ellis, I now feel so ashamed. We both, of course, had accepted the Marck. As time went on, Dan became more and more obsessed with his work. He even began to work on Saturdays, which, as you know, is not our tradition. We had so much, but his drive for more became obsessive. Not that I didn't succeed too, but when does one have enough? Gradually, our traditional Friday night Sabbath meal with friends faded and we stopped gathering with them and other acquaintances at Temple on Saturday mornings. When I did go, there was hardly anyone there. I guess other people were also preoccupied Slowly, I begin to feel more and more isolated.

Then one day a funny thing happened. My accountant, who was a delightful person, invited me to visit her congregation close to Ft. Lauderdale. It turned out to be one of those weird Messianic Jewish groups. But I started visiting regularly, mostly because I was welcomed and the folks were friendly. That's where I heard that this Marck thing was a sign of an end-of-the-age evil empire. I wasn't convinced, but began to worry.

Well, Dan laughed at me, but did not oppose my venturing out to visit the group. It was okay, as long as it did not interfere with our thriving business deals.

It was on such a Sabbath morning, when I was up in Fort Lauderdale, that the first comet hit the South Atlantic. When the 250-foot tidal wave crashed through Miami, I'm sure Dan was in his office on the fifty-eighth floor of the Dade Towers, a beach-front office complex,

Up in Broward County, our synagogue was located on the very western edge of Coral Springs, which was inland a few miles. In spite of the distance, the wave took everything, washing us all the way to the Everglades. As the rolling first wave crushed our little wooden building, I somehow caught a four-foot piece of the beam, lifting me to the surface. There I found myself atop wooden square, like a surfboard, riding a sixty-foot wave. I'm sure I was moving at least at thirty miles per hour. I was full of adrenaline and hanging on for dear life. Debris and bodies washed by and popped up and down.

It seemed like an eternity, but the wave gradually diminished over the next several miles until I could see the tops of trees. It was so scary, but finally -- Bang! I was tangled in a mangrove tree. That's where I landed, at last. I couldn't see any of the others who had been with me. I was on an Everglade hammock tangled in its branches. I guess that's what kept me from washing on across Florida. Thankfully, I was alive and conscious. Lot of bruises and cuts, but no fractures. Funny, my first thought was, well, that should take care of the gators and snakes for sure. That wasn't entirely true. There was a big snake, I'm sure it was a python, tangled in the same tree. He was probably too scared to attack, and quickly slivered off into the water.

Now, for the first time, I thought of Dan. Could he have survived? Emotions overwhelmed me. I wanted to believe he could and did, but knew there was no chance. I collected my thoughts and begin to think, *Now what? What do I do next?*. I thought if this was a tidal wave, the water will continue to subside. So, the best thing for me would be to wait it out in the tree. And there I stayed overnight and into the next day.

I could see that I was on an everglades mangrove island and climbed down to the swampland. At first, all I could do was to sit there and cry. I was there all that day and the next night.

Then on the following morning I heard voices. They were coming from another mangrove island a few hundred yards away.

"Hello, hello, hello."

Someone had spotted me. I stood and waved my arms, climbed back up the tree and yelled back. Bill and Margaret from the synagogue were over there, alive, and he had fashioned a mangrove branch raft. That's how I linked up with some others.

Over the next couple of days, we were able to find two more from the Messianic group, who had landed not too far from us. Now, unlike you, I'm no survivalist. I think roughing it is ringing for room service. But Bill had been an Eagle Scout and had some basic skills. We reinforced our raft, made wooded spears, fashioned mats and overhead cover from palm leaves. We worked our way north, beyond the Everglades toward dry land closer to Okeechobee. We were learning to live off the land and we made a plan to work our way further north and west. We expected there would be less damage on the west coast. We gathered canned goods, tools and water bottles from washed out stores and businesses. It was strange, we didn't see any other survivors for weeks.

And, my dear Ellis, it's still a long story after that. I'll bet you have some adventures to share too, but I'm exhausted and hope to get a little rest before dinner. I hope we'll have plenty of time to talk later, but for now your long-lost cousin needs to check into her room. Do you know where deck 2B. Room 322 is?

"Sure, I'll show you. Cousin Shulla, it's so good to see you! Are your companions on board too?"

"Yes, they went straight to their rooms."

DAY 17

In Miami, an Atlantic communications center has been established. Five more ships from Mexico, Central America, and the Caribbean join the convoy just out of Miami. They are led by another Kadosh named Sepharad. The Atlantic is noticeably rougher than had been the Mississippi or the Gulf. The hydrofoils now have to slow to thirty knots.

DAYS 18–21

Along the way, ten more boats from South America join the growing fleet along with more Kadoshim, Yehuda and Hannah. Some of the Millennium fleet boats from South America are over capacity, so a few of their passengers are assigned to the boats from North America, adding to the international flavor aboard the *Millennium Eagle*. Now every bed is occupied and at dinner every chair is taken.

DAY 22

Crossing the Atlantic takes six days. The big boats sight first landfall on the southwestern shore of Spain. Waiting off the coast of the port of Cadiz are nine more ships from northern and western Europe and the British Isles. The Kadoshim with them are Devorah and Yonah. Teah searches the rosters for possible Irish relatives. But there are none.

Isaac finds Teah and takes her over to starboard, to look at the Spanish coastline. Without hesitation, he rapidly starts talking to her about Cadiz.

"This is so exciting. Over there is the port of Cadiz. You know, it was founded by the Phoenicians all the way back in 1100 BC. Putting that in perspective, that was during the lifetime of the Prophet Samuel and Ruth and during the Philistine oppressions in Israel."

Isaac noted that he has Teah's attention, so he continues, "Cadiz was a center for trading tin and silver, from the British Islands to the rest of the known world. Those were the ships of Tarshish we hear so much about in the prophets' writings. The Phoenicians grew fabulously wealthy trading in these and other things, including spices and herbal drugs. Tin was absolutely essential for making weapons. Tin mixed with copper makes bronze, the chief hard metal of that age – the Bronze Age. Mastering the use of iron hadn't yet been developed to any large extent.

"The city of Cadiz was originally called Gadeira, which is quite interesting. Gadeira was a town in northern Galilee, a region which had been ruled by the Phoenicians during the same time. Some think the original Phoenician settlers of Gadeira named it after their home back east. The name Gadeira likely comes from the town name of Gad, one of the tribes of Israel. So were they Phoenicians, who moved west, or Israelites of Gad, who were sent west as colonists by their Phoenician masters? Maybe someday we will know.

"Gadeira became part of the Carthaginian Empire in 500 BC. That was about the time that Nehemiah, Ezra and Zerubbabel were rebuilding Jerusalem and the second Temple. The prophets Haggai and Zechariah were giving their prophecies then.

"The Carthaginians were Phoenicians anyway, so there was no change in language or culture, only to whom you paid tribute. The Phoenician alphabet is the same as the ancient Hebrew alphabet.

"When people talk about the Phoenician and Canaanite alphabets, they are really talking about the old Hebrew alphabet. That alphabet was adapted by the Greeks, and then passed to the Romans. It is the basis of our current English, Spanish, French, German, Romanian, and Italian alphabets.

"In the Punic Wars, between the Romans and the Carthaginians, the Romans took control. They renamed the town Gades, from a reference in the works of Herodotus.

"Skipping to more modern times, Columbus sailed from Cadiz on his second voyage to America. Sir Francis Drake burned the Spanish fleet in its harbor. During the Spanish golden era, Cadiz was the official center for new world trade. So here we are, part of the new world cargo, waiting to go home.

Teah leans over and gives him a big kiss. "You're so sweet. I just can't get enough of your stories. Keep them up." And she hugs him again.

Every one of the Rock River crew had noticed that Isaac was no longer lecturing them, but was going straight for Teah, who seemed to appreciate his lessons more than the rest.

DAY 23

The fleet of Millennium Big Boats lifts anchor and heads straight for the Pillars of Hercules in the Mediterranean Sea. There are now eight Kadoshim, Yoni, Natan, Chavah, Sepharad, Yehuda, Hannah, Devorah, and Yonah. They all address the fleet as it sails.

Devorah speaks for the group. "We have now gathered in this fleet, forty Millennium Big Boats. We are carrying forty thousand people, men, women, and children, all from the remnant of Israel. Others from the Mediterranean region have already gone on ahead of us. So we will not be joined by any more."

Ellis whispers to his team. "They probably just took their rafts!" A few smile, but most just look down, hoping no one heard him. Devorah, having heard, looks over at Ellis and shakes her head. But then she cannot suppress a slight smile.

She regains her posture and loudly proclaims her message to the whole fleet. "It is clear sailing now. Straight ahead. Next stop, Haifa, port city of Israel."

CHAPTER 12

MIDDLE SEA

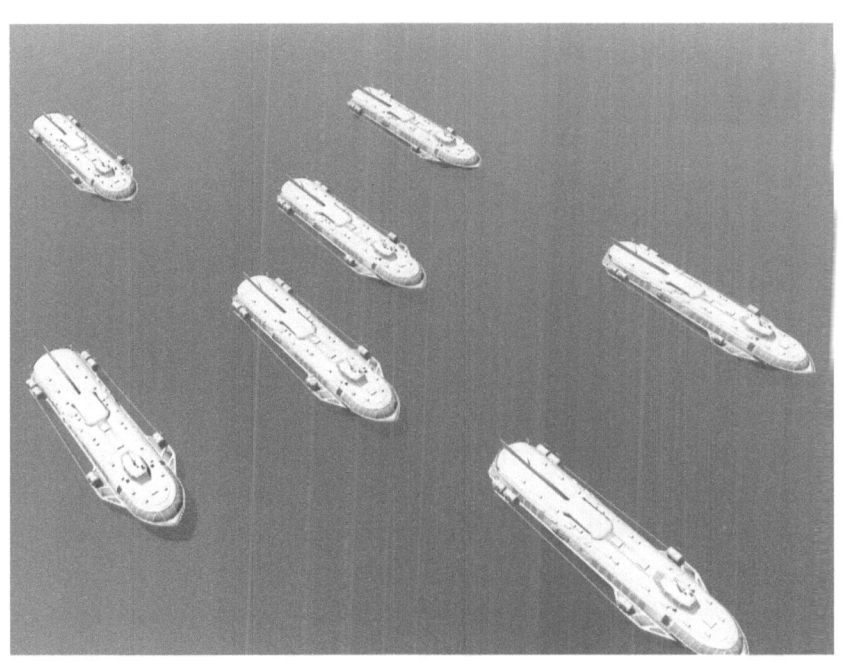

DAY 24

By midmorning, the Rock of Gibraltar is coming up on the horizon on the port side.

Isaac is overcome with the event of the forty millennium big boats sailing through the Pillars of Hercules. He pushes through the crowds, which are looking from the forward deck, until he finds Teah.

"Teah, those are the Pillars of Hercules!" Isaac is a little unsure about launching into another lecture. "You said you wanted to hear some more history?"

"Yes, of course." She relaxes, leans over on his shoulder and is ready for her next history lesson.

"Over here on the north side is the Rock of Gibraltar and over yonder to the south is Jebel Musa, the African pillar.

"The two mountains are associated with the westernmost travel of Hercules in his twelve labors; thus the name; but that's not conclusive. Some scholars think that the term used in ancient literature, 'Pillars of Hercules,' originally referred to the huge columns of the temple in Cadiz. That, of course, would put the Pillars of Hercules in Cadiz, Gadeira, which we saw yesterday. Just imagine, there're scholars who have dedicated thousands of hours of research to that debate. Where are the Pillars of Hercules? These mountains or the Temple in Gadeira?"

Teah asks, "Which do you think is right?"

"Well I'm not sure, but Plato seems to have favored these mountains. And, he recorded that Atlantis was beyond these mountains. Of course Atlantis had reportedly sunk so these great mountains, these pillars, to the ancient Greeks and Romans came to mean, 'nothing beyond.'

"That 'nothing beyond' was a great ocean which then took on the name of the mythological city, Atlantis. It became known as the Atlantic Ocean."

Teah says, "That designation 'nothing beyond' is very intimidating."

"Yes, it was to many, but the Phoenicians did not fear to sail beyond these mountains. They settled in Gadeira and traded for tin in Cornwall in southwest England. There's even some evidence of their travel and settlement in North America.

"Think about it, ancient great civilizations believing there was nothing beyond these pillars and here we come, in beautiful fusion-powered hydrofoils carrying 40,000 travelers, from that 'nothing beyond!'

The two, Isaac and Teah, along with the other travelers stop talking and for a long while just relish the moment and the view as the great ships sail through the Pillars of Hercules and into the Mediterranean Sea proper.

Teah says, "It is called the Mediterranean Sea after the custom of the Greek and later Roman names for Middle Sea. That is the sea in the middle of the lands. Even modern Hebrew uses that term HaYam HaTikhon which literally means 'the middle sea.'"

Isaac nods and smiles, "Didn't know that."

The sight of the fleet of gleaming white Millennium vessels with their silver reflecting windows sailing through the Mediterranean would have been spectacular had there been any spectators to see the sight. But they sailed alone. Well, almost alone.

With the straits barely four hours behind them, Ian, using his trusty scope, spots another fleet ahead on the horizon.

"Look!" He shouts, "There's a strange fleet of dark-hulled vessels ahead!"

The captain and the chief Navigator, Demetrios, each take turns looking through Ian's scope.

The captain says, "Yes, here it is coming up on the radar now . . . at the very edge of the screen. It's a fleet of about one hundred freighters.

"You see, we're not the only ones who are headed to Israel. Great fleets of cargo freighters are being assembled to bring in supplies, building materials, construction equipment, temporary housing, trucks, tractors, cranes, coal, oil, refined fuel, hydraulics, ores, pipes, plumbing, furniture, wood . . . you name it. They are bringing in what will be needed to build a nation again. Some also transport a few people along with the goods."

Ian asks, "Why is all that stuff needed?"

"You'll see when you get there. The destruction in Israel was even more devastating than it was in America and Europe. Most cities have been completely destroyed along with the smaller villages and suburbs. After the Israelis' quick withdrawal from the Union, following the Supreme Commander's proclamation that he is God, the Gia Union made good on its threats against them.

"Most has been destroyed?" Ian feels a flash of frustration and anxiety. Somehow he had assumed everything would be in order by the time they got to Israel.

"Yeah, in fact the only functioning port in Israel that's left is Haifa and it has limited capacities. So, it's likely that when we get there we may have to wait our turn, as the harbor and port facilities are expected to be filled with freighters unloading.

"The fleet ahead is slow moving, nothing like these hydrofoils. We'll catch up and overtake them within an hour.

"Go tell Ellis to send out a communication to all ships and to all passengers. They will need to know what we will be passing."

Ian runs down to Ellis' desk on deck three. He is really busy, but always glad to see his old friend.

"Ian, what's up?" Ellis says.

"Got an urgent memo from the captain. He wants you to send it out to all ships and all passengers."

"What? Let's see it."

"Well, I don't have it written down but here's what it has to say."

Aaron joins the discussions and quickly writes out the memo as Ian dictates to him. Ellis is listening intently.

"That's quite a message," Ellis says. "Seems like that is one the Kadoshim would want to make themselves. It has all kinds of implications."

Ian is impatient and not in the mood for a discussion. "Well, they're not here. We'll be overtaking a fleet of over a hundred dark ugly freighters within an hour. There's going to be a lot of confusion and hullabaloo if they don't have some sort of an answer or explanation.

"I think the Kadoshim want us to do what is right and we know they want us to take initiatives. They're obviously aware of this. I feel that the fact they have not appeared to do this is because they believe we can handle it ourselves."

Ellis pauses still. He hears steps behind him. He turns to look. "Yoni!"

Yoni has appeared behind Ellis. He says, "Go ahead and send the message. Ian is right." Then he disappears as rapidly as he appeared.

The message is immediately sent via the ham radio system to the other ship communication coordinators and captains. Ellis then turns on the ship's speaker, and the other ships patch him in.

He speaks into the microphone. "Shalom. Attention all crew. This is Ellis Coplan, *Millennium Eagle* and millennium fleet communications coordinator. I have a message from Captain Michaelpoupolos to everyone.

"Just ahead we will be passing a large fleet of cargo vessels. Although some of them may still have Gia Union markings, they are not Gia. They are friends. They are support ships going to the same place that we are headed.

"As we pass them, wave and send them your blessings. It's very likely you may soon be using some of the much needed restoration and rebuilding materials they are transporting for us and for Israel. The Kadoshim will provide us all more information later. Shalom."

Isaac and Teah are sitting on the forward deck, talking about what to expect in the coming days and weeks. They have already spotted the black dots on the distant horizon when they hear the announcement. They immediately run down the steps to deck three to Ellis' office. The office is chaos. The halls and doorway and office are packed with people wanting to know more.

Looking through the crowd, Ellis motions when he sees their heads. "Come on in."

Teah pushes through the crowd, allowing Isaac to then squeeze in behind her.

"Why didn't we know about this?" Teah asks. "Why is a massive infusion of aid and supplies needed in Israel? I thought that Yeshua and the Kadoshim had healed the land, had set everything right. I thought we had been through the worst, that we were headed for a land 'of milk and honey,' that everything was getting better day-by-day. What's really going on in Israel?"

Voices from the crowd that has gathered in and outside of Ellis' office seems to echo her sentiments.

Ellis is defensive and abrupt. "Hey guys, I'm just a communication coordinator. I, too, thought the message was a bit confusing. But Yoni himself ordered it sent. I'm just the messenger. Don't kill the messenger . . . our forefathers did enough of that. Besides, I only know what you know, I just sent the message."

Most disperse from deck three and crowds begin to assemble on the top deck and in the forward dining area, places with the best views to the front and the sides. As the Millennium fleet approaches the freighter fleet there is a general chatter pointing and questioning from person to person. The same scene is repeated on all the Millennium vessels.

The freighters are accompanied by small shuttle-craft and Gia Sky Pods which seem to be hovering above the fleet, as if to either protect or maybe to guide them.

Ian's response, "Well I'll be damned. Sky Pods!"

Isaac, Ian, Aaron, and Ellis are very familiar with the Sky Pods, having been trained as Sky Pod pilots years ago when they served in the Marines. The Pods were first built by Boeing in their Charlestown plant. Boeing had kept their new, revolutionary quantum-gravity technology a secret.

Isaac asks, "I wonder what happened to that super-secret Sky Pod technology when they were absorbed into the Gia Union Forces?"

Ian says, "Well those sweet little SOBs are still flying, even in the hands of a third party, so I guess we'll just have to wait and see."

While the Marines are focused on the Sky Pods off to the port side, the upper deck of one freighter is becoming clearly visible. Everyone on the *Millennium Eagle* can see as a door opens from an inner chamber and out walk about three dozen men and women. In addition, there are several children. They all look over and up at the *Millennium Eagle*. They begin to wave and smile and swing their arms as if to say "Hello."

One person lifts up a piece of blue cardboard with the word SHALOM printed on it. Several yell out. "It's written in Hebrew!"

Ellis now speaks on the ship intercom. "As you can see, there's no need for concern. These freighters, along with whatever else they are carrying, seem  to have on board refugees, remnants of Israel, the same as does the Millennium fleet. And those folks are freighter passengers. Unfortunately for them, their ships do not have the accommodations we have. In comparison, our somewhat spartan facilities suddenly now really seem quite luxurious."

Ellis decides to conclude with a lighthearted comment. "For one, I'm no longer complaining about having to go down the hallway to shower."

The familiar Passover song, "Next Year in Jerusalem" is now improvised by a group of Puerto Ricans to "This Year in Jerusalem." They sing the song over and over until eventually all are singing and dancing. Whatever is ahead, it will be enough. Dayenu.

Within an hour, the massive freighter fleet begins to fade into the west, as the much faster Millennium hydrofoil fleet passes them by.

It is now dinner-time. Most daily chores have been finished, and everyone is settled into the, now completely full, forward dining hall. Talk and discussion is minimal, considering the impact of the freighter fleet on the crew.

After dinner, it's Matt's turn to lead the after meal thanksgiving blessing. He pauses, then purposefully adds a blessing and thanks for the freighter fleet as he amends the old British navy hymn.

"Eternal Father, Strong to Save,
Whose arms doth bind the restless wave,
Who bidst the mighty ocean deep,
Its own appointed limits keep,
Oh, hear us when we cry to thee,
For those in peril on the sea."

Matt finishes and sits down. There is a moment of silence. Then, everyone in unison responds, "Amen."

The hydrofoils have opened up, full speed, on the smooth Mediterranean. Captain Michaelpoupolos and the other thirty-nine captains have decided to see what these boats can do. It is almost like a race. Some boats hit fifty-seven knots and all are averaging over fifty-two knots.

The discussions are somewhat quieter and more reflective that evening after dinner and in the lounges. Everyone now is more serious and contemplative. Upmost is what lies ahead. Doubts and anxieties arise. Most retire early to their quarters.

EASTERN MEDITERRANEAN SEA - DAY 25

Shortly after breakfast is served, Chavah appears at the front of the dining room. Her image and message is projected to the rest of the fleet.

"Shalom. As we sail nearer to our destination, it is fitting that we take some time to review and reflect on all that has happened. Our group comes from many diverse geographical and cultural traditions. Hence it is important that all of us have a correct and common understanding of what has taken place. I also want to give you some updates."

She has everyone's attention as she starts the briefing.

As I think everyone knows, the land of Israel, like the rest of the world has been under the tyrannical rule of Gog, Supreme Commander of the Gia Union and his henchmen. Through his first three and a half years of rule there was unprecedented worldwide prosperity enabling Gog to continue to tighten his grip. When he was convinced that he and his team were unstoppable, he committed the ultimate crime. He sat himself in the Holy Place of the Holy Tabernacle and proclaimed himself as God, Savior, King of kings and Lord of lords.

That triggered the beginning of the wrath of Yehovah, the real God in heaven, upon Gog and his kingdom. It is surprising and shocking that although the whole world knew that Gog had offended Yehovah, the God of heaven, most people still held to their unwavering allegiance to and faith in the Gia Union and its Supreme Commander, Gog. He had given them so much that most of mankind actually had begun to look on him and his government as the source of all good – as a god. HaSatan seemingly had now deceived the whole world, through Gog and his Gia Union. Remarkable as it seems, even though the series of increasingly severe plagues sent to shake his control and to awaken the world to his inevitable defeat, millions even stood in the firestorms and shook their fists toward heaven while still pledging allegiance to this Beast and his kingdom.

As the plagues escalated however, Gog's control over the planet began to unravel and there was growing discontent in many places. Ten-members of the European Union managed to muster a rebellion. But Gog brought his forces down on them and destroyed their major cities including their capital, Rome. The European Union then backed off and surrendered.

In America and Canada, rebellions were so frequent and so widespread that Gia Troopers went from town to town. They rounded up the population . . . murdering millions and taking millions more off to detention centers, modern-day concentration camps. Numerous others escaped and got away to the countryside. Many fled into Mexico, Central and South America, and the Caribbean. But America and Canada were effectively depopulated.

A Chinese-Indian-Central Asian confederation eventually sent two hundred million troops across the Euphrates toward Jerusalem only to be devastated by the Gia Air Forces using nuclear, chemical and biological weapons. That was only four months ago.

Toward the very end, another Balkan and Russian led Slavic rebellion was organized. To harshly crush that insurrection, Gog gathered whatever forces he had and he left Jerusalem to go north and personally assemble the striking force in the Jezreel Valley next to Har Megiddo. That is when Yeshua landed on the Mount of Olives and entered Jerusalem unopposed. I'm sure you all know the outcome of that.

Over this time, the majority of the world's population remained loyal to Gog, Gia Union Commander, even to the very end. The effect was a loss of ninety percent of the world's population, all directly related to war, famine, disease, and the plagues.

And of course, there's more. In addition to those four major rebellions, there were still millions of individuals and small groups all over the world who cried out to God asking for forgiveness for them and their families. And God has extended his protecting hand to those, just as He has toward you.

Now, I want to address a very important question that has been deliberated by most of you. Yes, there are the millions of the nations who Yehovah has saved alive, all over the world in every nation. They, too, are being cared for and organized. Right now there are ambassadors from Yeshua in their capitols assisting and guiding them. Those nations are part of the Kingdom and will join in the *rebuilding of the planet* each in its turn, *starting in Israel.*"

Chavah, finishes her review, but does not remain for questioning. She disappears out the back of the room.

These events are the ongoing focus of conversations after dinner, and in the forward lounge afterward until well into the night.

Miriam says, "That explains the absence of people in all those cities as our river crews drifted down the rivers. But I wonder where were they taken . . . and which ones survived? We all have friends and family still unaccounted for. I guess some are still alive somewhere."

There is an increasing desire to know more even though there is little this gathering can do about that right now.

That evening after dinner, several of the men from the Rock River group are sitting in the forward lounge. Tom and Bart, the two once missing Seventh-Day Baptists are sharing their experiences during their time in the Gia Union. Jacob, an elderly man from Brazil, comes over to join the group and introduces himself.

Isaac gives the old man a welcoming smile. "Jacob, tell us about yourself."

"I'm from Brazil and was reassigned to the *Millennium Eagle* from the *Millennium Lion*, which joined the convoy in the Atlantic. We were overcrowded and some of us were dispersed to other ships five days ago."

"Shalom and welcome," several say in unison.

"Are you originally from Brazil?" Ellis asks.

"Yes, I was born and grew up in Brazil but my folks came to Brazil in 1948 from Aleppo, Syria," Jacob says.

"Are you an Aleppo Jew?" Ellis asks

"Yes, the Jewish community from Aleppo scattered throughout the world after our community was destroyed in 1947. A large part of the community made its way to Brazil where we lived in peace until Gog made worldwide war on Jews just over three years ago--"

"Aleppo?" Isaac interrupts. "We know Aleppo. We fought there years ago, as United States Marines. We were sent in to protect the city from another genocide by its own government. I think the population had been part of a big rebellion many years ago . . . and in the Arab world, vengeance is not forgotten We spent eight weeks there.

"The town has an ancient and honorable history but was pretty much destroyed when we got there. I was already familiar with Aleppo because my uncle had a facsimile, a copy, of part of the Aleppo Crown, the *Aleppo Codex* — on display in his library. He used to tell me about its 1,200-year history.

"When we were serving in Aleppo, I was able to find the ruins of the ancient great synagogue and remember standing in those ruins thinking about how it must have been a hundred years ago."

Jacob says, "My grandfather, as a young man, was there when the Great Synagogue was destroyed. The family was quite well off and they were able to buy their way out of Syria after the riots. My father was just a child. They made their way, with several hundred other Aleppo Jews, to Brazil. where the family has lived ever since.

"Aleppo had been a large Jewish community for thousands of years. But in 1947, when the independence of Israel was announced, the Syrian government sanctioned mobs that stormed the Jewish community. My grandfather said the destruction was terrible as they raped, pillaged and destroyed the town. The great synagogue of Aleppo was burned and the Torah scrolls were incinerated. However, the Crown of Aleppo, a 1,200-year-old scroll of the entire Hebrew Bible was torn apart and scattered.

"The Jewish community, which had prospered in Aleppo for three thousand years, was dispersed. Over the next few years all the Jews fled."

Isaac says, "I learned a lot about the history of the Aleppo Codex from my uncle. It was written about AD 900, and was considered the most authoritative Masoretic text of the entire Hebrew Bible.

"Maimonides studied it. It passed from Tiberius to Jerusalem, brought by Karaite Jews. It was captured by the Crusaders in 1099, and had to be ransomed by Jews in Cairo. From there it went to Aleppo. It remained in Aleppo for over six hundred years, until those Arab riots in 1947.

"Like Jacob said, the *Aleppo Codex* was taken out of its safe and its pages were torn out and scattered. Over the next few days, some brave Jews risked going to the burned out synagogue and gathered as many pages of the Codex as possible.

"Later, Ben-Zvi and the Hebrew University put together efforts to rescue what could be found and eventually, in 1958, the remains were smuggled to Jerusalem. They were put on display in the Shrine of the Book in Jerusalem, along with the Dead Sea Scrolls. I wonder if they are still there.

"Actually, less than three hundred of its pages made it to Jerusalem, which left nearly two hundred pages missing. Fortunately, the Hebrew University was able to reconstruct the many of missing pages from notes taken from nineteenth-century scholars, and in 2010, published a printed version of the full text as the *Jerusalem Crown*. They also made a few copies of the existing pages, one of which my uncle the Rabbi had on display. But there has been an ongoing search for the two hundred missing pages of the *Aleppo Codex.*"

Jacob pauses, then reaches into his pocket and pulls out four folded, blackened pages of parchment. "Here are four of those pages."

Isaac is shocked. Speechless. Finally, he reaches out and carefully takes the pages, gently unfolding and admiring them. "Wow! Actual pages of the Crown of Aleppo! What can I say? How did you get these? You're not joking, pulling my leg or anything, are you?"

Jacob says, "Well, one day when you've got a lot of time and nothing to do, I'll gladly share the details, but I would rather hear about what you guys have gone through."

Isaac thinks Jacob may be a little uncomfortable about sharing his personal information. "I understand. We all have gone through so much, sometimes it's easier to just listen to others. So, when you're ready, let us know. But, I think there's a healing quality in sharing our experiences and you can't imagine how eager I am to hear more of your story."

"Really?"

"Oh, Yes. Four pages! Man, are you gonna be welcomed in Jerusalem."

"Well, actually we have more than four."

"What? Jacob, you've got to be kidding!" Isaac is astounded. "Sorry man, but right now I've got a lot of time and nothing to do, so if you're OK, I think we all will get a glass of wine and listen."

Jacob gets a big glass of fruit juice and starts his tale.

My grandfather smuggled these four pages out of Aleppo when his family bought their way to Brazil. He later learned that the Ben-Zvi Institute and the Hebrew University were trying to gather all these pages back to Jerusalem. But, we Aleppo Jews have for hundreds of years considered the Crown to be a mystic protector of our community. When it was torn apart, members of the community turned over many pages to the Israelis. But they kept back some as good luck charms, sorta like a hamsa. The good fortune aspect of the Crown has been a strong emotional component for our people.

My grandfather and four of his brothers and other relatives each kept a page or two folded in their pockets. They felt it provided protection from the evil one.

When the Israeli commandos attempted to assassinate the Gia Union Supreme Commander, Gog unleashed a genocide against Jewish communities, not only in Israel but worldwide, including Brazil. The Gia Union Troopers stormed into our shops, our schools, and our homes. We were well known to the government, as Aleppo Jews were business and community leaders.

My grandfather knew they were coming and he gathered us, his nephews and grandkids, and put us onto a boat going up the Amazon. He gave us a horde of gold and silver coins and then gave each of us three or four pages of the Crown. He said, 'May the Crown protect you, and may you someday take it up to Jerusalem.' There were forty-two of us, each with three or four pages of the Codex."

Isaac says, "Unreal! That's like over one hundred thirty pages!"

"Yeah it was."

We hid in an Amazon native village in the Andes near a source of the Amazon. Gia Troopers followed us up the Amazon, but none of us had The Marck and the Amazon natives with whom we stayed did not either, so they could not track us well.

After a while we were getting a little complacent. Then one day a unit of Gia Troopers showed up in our village. We hid while the natives welcomed them and treated them as honored guests. They actually invited them to dine for the evening meal. The following morning found them all dead, poisoned by the cuisine.

"We dressed like the natives did. We helped with the fishing, hunting, gathering and village chores. We just didn't eat some of the stuff that they thought was special. They thought that was strange and often laughed at us when we turned down roasted monkey or grilled rat. We became basically vegetarians, supplemented with some fish. In spite of what they thought about our crazy diet, they welcomed us and protected us.

Ellis asks, "Why'd they receive you all so well? I mean, it was a lot of you to feed and protect, putting their own lives and families at risk?"

"Well, we had a kind of relationship with them. When my grandfather and his brothers came to Brazil, they heard of a tribe on the slopes of the Andes where the adults developed a kind of paralysis and died in their thirties. Uncle Raphael was a physician in Aleppo. He took a couple of trips to those Amazon villages where this condition was reported and he studied the people. He learned their language. He became friends with them and traded stuff from the coast with them. He noted the middle-aged adults stumbling and loosing balance. And there were no older adults.

Uncle Raphael had a respected infectious disease practice in Sao Paulo. Also, he was following the work of Blumberg and Gajdusek who found in the 1960s that New Guinea tribes had a disease called Kuru, caused from eating the brains of their ancestors as a part of their funeral rites. They were able to get brain samples and take them back to the United States where they injected them into monkeys. Two years later the monkeys developed Kuru.

This kind of delayed neurological disease, an infection, was originally thought to be a 'slow virus.' Now it is known that it is from an abnormal protein called a *prion*. Gajdusek and Blumberg got the Nobel Price for this work. But more importantly, the New Guinea government put an end to the tradition of eating brains and the disease was almost eliminated.

These prion diseases were later found to cause scrapie in sheep, mad cow disease in other animals and people, and Creutzfeldt-Jakob a disease in humans from infected corneal transplants.

So Uncle Raphael, watching all this unfold, thought it was time to visit his friends high in the Amazon. When he arrived, many remembered him, but many of the older folk were now gone.

He spent some time living with them, getting to know them again for several weeks. Sure enough, when a thirty-eight-year-old woman died, the tribe sat around and sampled her brain. It was to them as if they were eating her spirit, to live on in them. He did not scold or pressure them, but over the next several days begin to discuss what had happened to a similar tribe in the mountains of New Guinea."

Everyone listed with interest, but with skepticism.

Finally, the chief spoke up, saying, 'Raphael is our friend, he comes to give to us and he never takes from us. I think he wants us to live and be prosperous. Raphael, if we stop eating the brains of our ancestors how long will it be before we can tell if you are telling us the truth or lying?'

My uncle, not knowing the real answer, tried a theory, 'Currently, you all do not allow your people to eat of the brain, the spirit of your ancestor, until they are thirty years old. It is reserved for the elders. So, I propose that you not let the younger eat until they are forty. I will return in ten years, and we shall see.'

But the chief was skeptical, 'Few of us live for ten years beyond age thirty. If you are right, we will have more of us older tribesmen when you return. If not, you will have some explaining to those denied a part of the spirit of their ancestors.'

That was in 1969. And good to his word, Uncle Raphael returned, with three other physician friends in 1979. There were no new cases of the disease in those who were below thirty-years-old ten years before. And those that had been older than thirty were gone, including the wise chief. The new chief and council was convinced. They issued an edict that eating or even touching the brains of dead ancestors was forevermore forbidden. And the people never forgot what Uncle Raphael had done for them. He continued to visit them every few years. He took me with him on his last visit, the year before he died at age ninety-four.

So when I showed up with my cousins three years ago, the tribe considered us to be brothers.

We were there with them through the plagues and firestorms and earthquakes. So monkey, roach and rat became less common anyways. We all had lots of roots and berries, fish and ground bark.

We were there until Yehuda, our Kadosh, came to gather us.

Isaac was speechless and had to pause and take some deep breaths, "What about the natives?"

"They were invited to join us on our trip down the Amazon and to come to Israel. We couldn't just leave them. In fact, they had good canoes.

"One hundred and seven Amazon natives are aboard the *Millennium Lion* now, with my cousins. A few of us overflowed here to the *Millennium Eagle*."

Isaac says, "Thanks for sharing with us, Jacob. We've added another interesting story to a log we are keeping. And importantly, the Aleppo Crown is also being re-gathered too! My Uncle Eber would be so pleased."

The group lingers in the forward lounge talking and sipping wine during the evening. Then a message comes over the speakers. Ellis is sending a message to all ships, for all hands.

"Everyone is requested to return to their quarters early this evening. Get a shower and pack your bags before retiring tonight. Tomorrow will be a big day. A really big day. And we'll start early."

DAY 26

Music awakens everyone early the next morning. A voice comes over the intercom ordering, "All hands on deck. All crews are to assemble on the top and forward decks."

It seemed too early, but everyone complies.

Chavah is on deck already. She addresses everyone. "I thought you would want to see sunrise over Haifa. Welcome home!"

Many exclaim, "Wow, see that? Oh wow!"

Raquel gives Ellis a big bear hug.

Teah grabs Isaac's hand.

Everyone stands there in awe!

The morning sky outlines Mount Carmel in the background and the still dark harbor is filled with giant freighters.

The forty Millennium big boats then pull to the north of the port facilities where many freighters are vying for limited moorings. It appears that rather than using tender boats, the unloading is being done by Sky Pods. There are several dozens of them silently sweeping overhead toward the shore. Some people from the freighters are already on their way. The Millennium fleet drops anchor about one thousand feet off the coast, a mile north of the port moorings.

Anchors have not all been dropped when a Sky Pod approaches the *Millennium Eagle*. It hovers over the forward deck and settles down for landing.

Yoni, Natan, and Chavah appear together on forward deck.

"This is our transport to the city. What did you think? We were going to make you swim, or paddle, or build another raft?" Yoni says with his customary smile and wink.

"Gather your bags and meet in your assigned designated areas.

"Each Sky Pod can shuttle a dozen people with gear. The run to the shore is five minutes. Load and unload as efficiently as possible, because we plan on a run every fifteen minutes. With forty thousand hands to transport, using one hundred Sky Pods, it will be taking us eight to ten hours to carry everyone, with no breaks.

"Teah, we ask you to take the last one to make sure everything has gone as planned and the organization is running smoothly, ship side."

"*B'seder,* okay," she replies.

Isaac quickly asks, "Can I stay to help her?"

Chavah says, "I thought you would ask Isaac, but we need you, Ellis, Aaron and Ian to pilot some of the Sky Pods for us. We have one hundred Sky Pods, but we are short on pilots. I know you four used to pilot them back when you were in the Marines."

"Yes ma'am!" Isaac, Ellis, Aaron and Ian snap to attention.

To pick up their own Pods, the four, bags already packed, are on the first Sky Pod group to shore.

Teah is busy making sure everyone has their shuttle numbers and times and help, if they need any further assistance or have questions.

CHAPTER 13

HAIFA

HAIFA - DAY 26

This is the first large group of settlers to arrive in Israel. A few hundred have come in over the last two days, and a few dozens more came earlier as passengers on the first round of freighters. They are all busy with advance work.

But what is immediately evident upon their arrivals is that there is no housing for the forty thousand new pioneers. The port facilities occupy much of the harbor, and much of that is in disrepair and cluttered. However, there is a broad plain extending from the Mount Carmel chain to the east and south and bordered by the hills of Acre on the north. The best camp spots are south and east of the port and to the north toward Acre.

Natan addresses the arriving settlers. "A decision was made before the fleet's arrival to establish the first camp to the south and east. Land has already been cleared south of Ahi Eilat Garden up to the northwest slope of Mount Carmel. Several freshwater reservoirs are adjacent to the camp further to the east. We have dubbed the site Sukkot Gardens.

"The plan is to get everyone on shore to the new Sukkot Gardens camp as soon as possible, even as the camp is under construction. Assignments for individual and family campsites are already underway by the advanced ground crews."

The four Marines quickly locate their Sky Pods.

Ten minutes of flight review is all it takes. The controls have not changed over the last seven years.

"Ooo rah!" they yell as the Pods lift off slowly. They then accelerate northward. They swing around the fleet and out over the sea, getting a feel of the controls and Pod capabilities.

Yoni comes alongside the four Pods. "Hey, guys," he says, "This is a work party, not a joy ride."

"Just checking these babies out," Isaac replies, "Ooo rah!" as he does a figure-eight loop, then a 360-degree roll. The other three follow his pattern. Then all settle down and come to a halt, hovering thirty feet over the forward deck of the *Millennium Eagle*. These four Sky Pods are joined by six more. They all settle down onto the deck together.

"All down," Ian says through the speaker. "Let's load up."

Ten to twelve people, with their bags, crowd into each sky pod and they then gently lift off. At first Ellis leads the way.

He says, "Okay. Let's see how these flyin' machines handle fully loaded." The Pods lift without any apparent drag from the extra loads. They tilt slightly to the right, then rotate toward the shore and quietly glide into the distance, toward the new Sukkot Gardens camp.

Aaron, now leading the ten Sky Pods toward the camp, calls to the others. "Guys, where do we land?"

No one has actually said where to drop off the passengers.

Isaac is the only one to reply. "Let's make a landing zone on the northwest side, nearest the port. I think the supplies will come from the west, from the ships in port."

All concur. As the Pods settle to the ground on the northwest border of the camp, small John Deere Gator tractor trucks are already bringing equipment from the ships at port. All the other Pods and trucks converge on the selected staging area.

Over the next seven hours, the pilots rotate with the other craft, picking up passengers from the forty Millennium ships. Even while this is underway, word is received that another fifteen Millennium ships will soon arrive. Coming up through the Suez Canal from East Africa and India, they are expected late that night or tomorrow morning. Word is that even more will be arriving over the days ahead.

And as for the forty-vessel fleet, even while the unloading is underway, the ships are being restocked. The plan is to pull out tonight for their next mission. It looks like the Rock River Marine team will have a flying job for some time.

By mid-afternoon, all forty thousand have been transported to their new Israel home in Haifa, in Sukkot Gardens. The camp is intended to eventually hold 150,000 refugees. Isaac circles over and around the camp perimeter, watching the busy movement of people and supplies throughout the camp. They first settle to the south and east, making room for additional settlers to add their sites toward the west as they arrive.

Yoni finds Al and asks him to assume a major task. "Al, I understand you have experience in commercial plumbing. Can you take charge of running water pipes from the reservoirs to key spots throughout the camp for washing, bathing, and drinking?"

"Yes sir," Al replies.

"Let's start with selecting the sites for the bathrooms, kitchens and drinking fountains. We should begin with the eastern camp where our current settlers are getting established. Over the next few days we will expand towards the west and north. We may need some auxiliary storage tanks toward the westward camp."

Al surveys the arriving settlers and selects volunteers for the plumbing teams. The reservoirs are uphill from the camp, allowing water to flow to the facilities without pumping. Trucks with PVC pipes, filters, couplers, T-joints, sealant and plumbing tools are already arriving on the banks of the reservoirs.

Al assembles his group and enthusiastically informs them of the plan. "Team, our goal is to have running water for much of the camp by dark."

"Aye, aye, sir," the team responds.

Along with others, Michael and Joel are walking about the living quarters area, looking puzzled. "What are we building here? Anyone have a plan?"

Supplies keep coming from the freighter fleet, delivered by the Gator tractor trucks. Everyone is being supplied with 2 x 4s for framing, nails and rope, plywood for side walls, and woven grass mats to serve as roofs.

Miriam says, "Give me some palm branches, some decorative fruit and flowers, a lulav and etrog, and we will have a sukkah!"

"You got it," Natan says. "The festival of Sukkot officially ended last week. But Yeshua has officially extended it, like in the days of Solomon and of Josiah. Welcome to the biggest Sukkot Festival you have ever seen. And you will get your lulav and etrog. You will just have to share them with a lot of people."

Many of the people have never seen a sukkah before, let alone built one. Miriam, Rivka and Raquel at first, then Michael and Matt, circulate around giving tips and instructions.

Trucks keep arriving with more supplies, tools, wood, mats, some folding tables and even chairs. Six larger trucks arrive, each with a dozen Big John porta-potties.

"Don't worry," yells the driver, "there's a boat load of more to come!" He enjoys the statement, thinking he is really funny, and drives off with a big smile.

More trucks arrive with portable showers, functioning from overhead water cisterns. Several others arrive with big canvas tents, to be used for dining facilities, meeting and worship halls.

The plumbers, now ten crews, are busy running pipe from the nearby water reservoirs to the dining, bathing and washing facilities strategically placed throughout the camp.

As the evening sky begins to darken, Raquel sits down in her folding chair outside her. "I can't believe we got all this set up before dark! Good job team!" She takes a deep breath and leans back. Indeed, by night-fall everyone is settling into his or her own personal or family.

Yoni arrives loaded down with a couple dozen boxes.

"Shalom," Raquel says, "what's in the boxes?"

He says, "Lulav sets. Here is your lulav and etrog, *Chag Sahmayach*, Happy Feast!" She takes an etrog from one of the boxes, smiles and then gives him a big bear hug.

"We have a few boxes to distribute tonight at dinner, but I wanted to bring you yours first, since you asked for one."

"Now, Yoni, I didn't actually ask for one. I just said, since we are going to be living in these sukkot, if you give me a lulav and etrog it would be Sukkot for me."

"Same difference. Here's yours. Enjoy!" He lifts off with his boxes for others.

"*Todah,* Thanks!" Raquel yells.

Teah arrives from the Big Boats only an hour before dark. She has surveyed the Boats to ensure everyone has disembarked and made it to shore. Knowing she would be last to arrive, Michael and Miriam have set up her sukkah. She is personally touched by their thoughtfulness. She gives them both a big bear-hug.

Suddenly, Isaac shows up at her doorway. "Hey guys, want to take a ride and see something special?" he says.

"Sure, let's go!"

The Marines jump into their Sky Pods with their invited passengers, link their audio speaker units, and zoom up and over the camp, sweeping back and forth. They then rise over the slopes of Mount Carmel and along its ridge. Once over on the other side of the mountain range, they can see into the Jezreel Valley. It is littered with wrecked military vehicles, many still smoldering. The four Pods slowly ease down toward the valley, surveying the carnage. They can see not only the wrecked vehicles, tanks, planes and missiles, but also they can make out dismembered human bodies with Gia Union uniforms.

"Let's not go over there right now, Isaac," Teah asks. "That's for another day, Okay?"

The pods skim back north and east, and then along the mountain crest as it drops into the sea. They swing back east over the Sukkot Gardens below. There they come to a mid-air stop and hover, and admire the scene.

Teah says, "It's so spartan, so rustic, yet in some way so beautiful. I don't know how long we will be here, but it already seems like home."

Isaac says, "Yoni says they are expecting to house 150,000 settlers in this camp. In a few days or weeks, it will be four times this size."

Miriam says, "Hey Aaron and Ian, I hope that someone explained that they call it Sukkot Gardens because sukkots are temporary dwellings."

"Yeah, Isaac explained."

"Good. You know that they are meant to remind us of our fathers residing in temporary dwellings for forty years before entering into the land of Canaan."

"Well Miriam, let's hope we'll not be here for forty years," Ellis adds.

"I don't think so. Unless we rebel like our forefather's did. Let's keep our chins up and attitudes positive."

"We agree," say Aaron and Ian together.

The four Sky Pods continue to hover over the camp, watching the sun sink below the horizon of the Mediterranean Sea to the west. Teah leans on Isaac's shoulder as they glide back to the ground on the north side of the camp, just in time for the shofar call for dinner.

In the dining tents, teams of servers already have plates set and food arranged on the long folding tables. The teams that have gotten to know each other over the last two weeks are grouped together, with a few new faces here and there.

A blessing is said, the meal is served and a thanksgiving given afterward. Then Yoni, Natan and Sepharad appear up front.

Yoni says, "*Todah rabah*, Thank you. For all your good work, enthusiasm and service to each other. We are due to have another 10,000 join us tomorrow. This camp is scheduled to hold 150,000, maximum, all going well. We have arranged for one other such camp to be set up within a couple of weeks.

We are initially bringing 300,000 through this Haifa facility. You have gotten off to a good start. Thank you again.

"However, we will not be here very long, so don't get too comfortable." Yoni chuckles and then does his famous wink, wink.

"Pending how our sky pod unloading system works and after the harbors at Eilat, Joppa, and Caesarea are upgraded, more camps will be established. We want to bring back to Israel up to two million this year, maybe more. You are the first of those. You have the privilege of being pace setters and building a foundation for others."

Sepharad says, "From here, over the next few weeks, we will take breaks in our chores to go up to visit Jerusalem. We will go in shifts to meet Yeshua and many, many others you know about, or will come to know about.

"Soon you will be given your own place, your own land, your own position within the Kingdom of God on earth. As you can see, these are very temporary dwellings. That is very meaning of the word *sukkot.*

"For any who do not yet know, sukkah is the single form for such a dwelling, the pleural form is sukkot. We have some fruit and flowers and some artwork for some who may want to decorate their sukkah. And we have lulavim and etrogs. Miriam can arrange this with you.

It is Natan's turn to speak. "Now, beyond accommodating our growing camp population and the logistics of keeping it clean, safe, pleasant and efficient, we have some other pressing needs beyond Haifa.

"We need several thousand volunteers to help us clear the roads and to bury the dead from the recent battle of Gog and Har Megiddo. It is just the other side of this mountain range and to the south toward Jerusalem. It will be dirty, depressing and hard work. So no assignments have been made. Volunteers only. An ancient estimate is that it may take up to seven months to clear all this out and cleanse that land."

Ellis asks, "What ancient estimates?"

Natan says, "Both Isaiah and Micah foresaw this time thousands of years ago. 'And they shall beat their swords into plowshares, and their spears into pruning hooks: nation shall not lift up sword against nation, neither shall they learn war anymore.' Isaiah 2:4 and Micah 4:3.

"Wars are over. It's time to change the implements and resources of war into implements of peace and justice. But as we stand here now, the results of war, the biggest and last war have defiled the land, the land of Israel. This land will need some time and a lot of effort to be cleansed from the millions of rotting corpses, the burned out vehicles, crashed planes and rockets and the implements of war.

"Yes, Ellis, this too was foreseen by the prophet Ezekiel twenty-six hundred years ago. He even predicted the period of seven months needed to clear and cleanse the land. That prophecy said, 'And there shall they bury Gog and all his multitude, and they shall call it the valley of Hamon-gog. And seven months shall the house of Israel be burying of them.' Ezekiel 39:11–14.

"The prophet made the prediction, but it is left to us to fulfill it, to do the work. It will be a hard, dirty and disgusting job. But it is the first challenge. Volunteers only. Teah and her team already have signup rosters.

"Thank you, shalom, and *lei-la tov,* good night."

Everyone drifts back to his or her sukkah. Many begin to decorate them. Some don't know how. It is very noisy, as there is so much visiting and chatting long into the night and the buildings are in close quarters.

But Isaac, though very tired, sleeps poorly.

HAIFA – DAY 27

The next day at breakfast, Teah announces that the initial Jezreel Valley clearing and burying crew rosters are full. "We cannot take anymore volunteer names for now, as more have already signed up than are needed. But since the work shifts will be only one week and the groups will rotate, there will be opportunity for all. Thank you all."

Two thousand men and women are shuttled to work sites by the Sky Pods throughout the day. They will have a one-week shift. Then another group will rotate in. They will camp out at the work site and move, on their own, as local clearings are finished. Each person has a week's supply of food and water.

They have no equipment to actually repair the roads, only to clear them from obstacles, large rocks, burned-out tanks and personnel carriers, crashed helicopters and such. The burial detail is especially gruesome as most of the bodies are in advanced stages of decay, having served as carrion for vultures and wild animals, and of course, lots of maggots. The bodies of this army are to later receive their final burial in a cemetery especially designated for them, the valley of Hamon-gog. David referred to this place as "the valley of the shadow of death," in his famous twenty-third psalm over three thousand years ago.

Until transport is available, the plan is to bury them in mass graves off the sides of the roads, with site markers. There they will return to the dust from whence they came. Their bones will later be gathered and placed in the valley of Hamon-gog as a memorial for all to see and consider.

DAY 28

Teah is keeping track of the influx of settlers. She notes in her record book, "Sukkot Gardens is growing daily. It now has a projected population of eighty-nine thousand by the end of our first week. It seems that the first fleet of forty vessels was the largest. Now the fleets, arriving almost daily, are only ten to twelve each."

Isaac has heard of a discovery. "Guys, I was told there is an old, battered Torah scroll on one of the freighters. I'm sure I can get ahold of it, if we want it."

Miriam is very excited about the possibility. "If Sukkot has been extended, and yesterday was the end, then today is Simchat Torah. That's the annual time for the re-rolling of the Torah, and the rejoicing in the Torah." "Let's do it," many say.

Isaac immediately takes off to retrieve the old scroll and Miriam quickly arranges for the event.

Plans go well and it is not long before, amidst singing, dancing and praise, the final verses of Deuteronomy are read and the scroll is rerolled back to Genesis, where the first verses of Genesis are read. The annual Torah reading cycle is restarting on today, Simchat Torah.

DAY 33

Despite the hustle-bustle of the crowded coastal plain of the port city, and the large Sukkot Garden camp, it is quite dark after sunset. There are no campfires, due to the shortage of wood. And there are no street lights or house lights. There is no electrical system operational. Vehicles generally do not travel at night and there is really no need for police or army patrols.

Teah has been busy with her administrative and coordination duties and so has Isaac with his transportation duties. They have not seen each other in several days. After the daily chores and assignments are done, Isaac finds Teah to take her for an evening walk.

She enthusiastically agrees. "That's what I need. Thanks for remembering me."

They find themselves on a winding path going up the western slope of Mount Carmel. They stop and look back over the city, the harbor and the coastal plain filled with sukkot and people. The view stretches out into the Mediterranean Sea.

They sit on a rocky ledge and do not speak for a long time, looking out over the awesome scene.

Teah says softly, "Blessed are you, Yehovah, the God of Abraham, the God of Isaac and the God of Jacob, who has in his everlasting faithfulness brought us through the net of the adversary, who has looked down on His people and has so marvelously brought home, his people Israel. *Todah raba,* Thank you very much."

Isaac says, "Amen."

As they sit there in wonder, a thin crescent moon appears in the west as the sky darkens.

Teah says, "The stars are so bright, it is as if they are pressing down upon us. I feel as if I could just reach out and touch them."

Teah reaches for Isaac's hand and points upward with her other. "Tell me, what's that star?" She leans over against him.

Isaac rests back against the rock and begins, in his lecture mode. "That's not a star, it's an open cluster. It's one of several dozens in the region of Sagittarius.

"The constellation is called Sagittarius, a centaur, half-man, half-horse and an archer. Most have trouble imaging a centaur-archer up there, and we generally refer to it as the 'tea pot.'

"See the four stars forming the body of the kettle, then see the spout and lid? And then the two stars making a handle? See?

"If we had some binoculars, we could explore dozens of open clusters in this constellation. And you can see that this is right in the middle of the Milky Way, in fact, if you sweep your eyes back and forth along the Milky Way you can see it bulges, getting wider here in Sagittarius.

"You can even see a bright glow around the Milky Way here. What many people fail to realize is that, from way out here from the edge of the Orion spiral arm of the Milky Way, we can still look at the center of our galaxy. The galactic center is only blocked from direct view by the thick dust clouds of the Milky Way's spiral arms. Here looking toward the galactic center, we are looking through three spiral arms."

Teah says, "Isn't that where the giant black hole is, where HaSatan was thrown into nearly three weeks ago?"

"Exactly, the hologram-scroll, projected for us in the sky, showed him being bound in light chains and then hurled through a worm-hole that leads to the galactic central giant black hole. So I guess that is where he is now, if he actually can exist in such a place.

"There are giant black holes in the center of every galaxy. They are essential for the formation of galaxies. There is a relationship between the velocity of the stars in the outer arms of the galaxy and the size of the black hole. There is also a direct relationship between the size of the galaxy and its central black hole.

"Enough about black holes and star velocities. Teah, it is so good to be with you again." He shyly leans over and gives her a shoulder hug. "Tell me about you. How are you doing?"

CHAPTER 14

UP TO JERUSALEM

HAIFA – DAY 59

Teah remains occupied with her duties, associated record keeping and reports. Sukkot Garden is now 145,000 strong. A second camp is being set up just south of the citadel of Acre. This is four miles north of Ahi Eilat Garden and Sukkot Garden. The residents of Sukkot Garden are by now quite experienced and confident in sukkah living. They will be serving as experts on the ground for those who start arriving at Acre Beach City later today. That will require daily travel to help the newcomers adjust and problem solve. Settlers are still coming, but their pace is slower and steadier.

Isaac remains busy shuttling the newcomers to their new homes and, in addition, providing shuttling services as the road clearing crews move back and forth.

At breakfast he shares his feelings with Teah. "I'm beginning to feel a lot like a bus driver. But, still every day is an adventure and I have to admit that it's really a cool bus. And as we shuttle the work teams throughout Israel, I'm very interested in seeing and learning the land from my vantage in the sky pod. I am even beginning to look at spots where one day I may like to homestead . . . well maybe.

Teah says, "Oh Isaac, this is all so exciting. I agree some of our work is tedious, but when you realize how much is being accomplished, it just makes me feel good all over. And the thought of someday being settled in a home of my own . . . it's wonderful."

"Well, there are some beautiful areas out there, once you get by all the destruction and carnage. I'll sure be glad when we get past these clean-up efforts."

Teah says, "Yeah, that work is really overwhelming, but you rarely hear any complaints from the crews."

"You're right. Yesterday, I shuttled the fourth road clearing and burial rotation group back to Sukkot Gardens. You know, in spite of the fact that the Mediterranean Sea is now getting cool, the workers wanted to be dropped off on the beach, so I agreed. Some other Pods joined us and the crews spent over an hour soaking in the water, playing in the waves and generally cleaning up before we were to return-shuttle them back to camp."

Teah says, "That must have been fun and therapeutic for them. After what they are going through, well, I'm sure they needed to clean their bodies and their mental state."

"Yeah, while they were in the ocean splashing in the cool waters, I went back to camp and brought them towels and clean clothes. I kept thinking about how everyone was being so stimulated and invigorated by the seventy-five-degree water. You know, growing up in Wisconsin did not make me immune to cold ocean water, so I parked the sky pod and took a dip myself. Heck, Lake Michigan never got that warm."

"Good for you! Every bus driver is entitled to a holiday."

"Very funny. Hey, while at the beach I saw Tom and Bart, you know, the two missing Seventh-Day Baptists. They were part of the crew so I got a chance to spend some time with them. Teah, I'm a little concerned about those two. They both seemed uncomfortable and expressed concerns about fitting in. I tried to reassure them, but I don't think it did any good. When I get a chance, I think I'll mention it to Yoni."

"Remember Isaac, those guys are experiencing some major cultural changes and their strict upbringing has probably not prepared them for all this. I'm sure they may be concerned about how and where they fit in. But, they did volunteer for the cleanup efforts, so I think they will be Okay."

"That's a good point, but I'm still going to mention it to Yoni. But, for now, it's time to get my Sky Pod and be off for another adventure! You take care of yourself too."

By evening, Isaac has still not stopped thinking of Teah. He finds her again and joins her at her table for dinner. "Two meals with you in one day. That's great! Sometimes I don't get to see you for several days in a row," he says.

They are chatting after dinner, just generally catching up. Teah seems to know everything, and indeed she does, as chief activity coordinator. Suddenly they hear a clearing of a throat behind them. Looking around, Yoni has dropped by to visit.

He starts with an unexpected and blunt question. "So, folks, is it time to go to Jerusalem and visit Yeshua and friends . . . or not?"

The impact of the question shocks Teah and rattles her a little. "What? What'll I wear?"

Yoni is amused at her response. "Okay, you decide. Pack for a three-day trip. We leave the first thing in the morning. We will go by sky pod. Isaac, I want you to pilot one of them. We will take fifty pods carrying six hundred people. The road clearing and burial volunteer veterans will have first crack at it. Eventually everyone will go."

Isaac realizes an opportunity and interjects, "Yoni, Tom and Bart have already been on the cleanup details. Do you think they might be included in this group? They seem to be a little frustrated and appear to be having some trouble sorting this all out."

Yoni grasped the significance of Isaac's suggestion. "Of course, I'll make sure they are on the roster.

"And Teah, you should also come along for the first trip. Maybe you can help us arrange the future visits more efficiently. We will soon be getting some new Sky Pods and we can move toward enlarging the visits. Want to hear your ideas."

"Yes, I'll be happy to do what I can."

"Oh, and just wear whatever you think would be appropriate, when you are going to meet the King of the world!" He smiles and does his usual wink, wink. It's his sign when he is playing with your mind.

And he is off, announcing the same to five hundred ninety-eight other hard working settlers ready for a break.

DAY 60

Morning comes too fast. Teah had problems sleeping too. But choosing clothing wasn't so tough since there were only seven choices, and they all looked about the same.

Fifty Sky Pods are ready and take off right after a quick breakfast. Yoni is on the speaker talking to all the Sky Pods while they are traveling. He takes on the role of a tour guide pointing out things of interest or things that should be of interest, at least.

If you look off just ahead to your left, you can see the high ridge of the city of Nazareth. It was very small in the days of Yeshua, only two to three hundred people. It grew to a major city in the late twentieth century, covering the entire mountain. But it was destroyed in the Israeli revolt against the Gia Union three years ago. Upper Galilee has always been known for its resistance to oppression and has had its share of radicals over the centuries. Nazareth comes from the Hebrew word, *Netzer*. Netzer means branch. It is said to have received that name because so many of its small population were of royal or priestly blood. There were many branches of the royal and priestly line living in that small village.

In the distance beyond Nazareth, off more to the left, is the famous Mount Tabor. I can just see Deborah, with her assembled army of ten thousand Naphtali and Zebulun shock troops, suddenly descending down the slope to surprise Sisera and the Canaanites below. On its northeast slope is the village of Endor where Saul went to enquire from a witch. The name comes from Ein (spring) of Dor. There was once a kibbutz there before the Gia Union destroyed it.

Mount Tabor rises in the middle of the famous Jezreel Valley, which we all know too well, especially for those on the burial duties. It swings east to west across from the Jordan Valley all the way to the Mediterranean Sea. It was famous for its agricultural output but now you can still see the thousands of burned-out tanks, helicopter wreckage, ruins of missile launchers, and all kinds of implements of war. A lot of work remains to clean up all this mess.

Off to the right is the hill of Har-Megiddo. Look, the Gia Command Center has survived and its silver dome-shape still shines in the morning sunshine.

Moving south into the hills of Ephraim, to your right, are the ruins of the city of Samaria, the capital of the Northern Kingdom of Israel. From there war was waged by Israelites upon Jews in the Kingdom of Judah. Brother against brother. Finally, it ended in 722 BCE when the Assyrians conquered it. They took the Israelites of the northern kingdom into captivity in Assyria and moved people from Babylon into the area. The new stock became to be known as Samaritans. They opposed the restored Kingdom of Judah and did all they could to harass the Jews.

"Thus, by the time of Yeshua, Samaritans were despised. That is why there was so much controversy generated by Yeshua's conversation with a Samaritan woman at the well, his depiction of the Good Samaritan as a good guy and his traveling right through Samaria, instead of walking around it, to get from Galilee to Jerusalem. Samaria is still in ruins.

As we see a bit further to the south, on the left you can see the twin peaks of Mount Ebal and Mount Gerizim. On the slopes of those mountains, in the day of Joshua, the Levites put representatives from half of Israel on Mount Ebal and the other half on Mount Gerizim. They then read the blessings and cursings. Israel responded by 'Amen' and pledged to keep the Torah. We know how that turned out.

Between the hills is the town of Shechem. There Abraham came to the great tree of Moreh and built an altar where he offered sacrifices to God.

As we pass further south, we will fly over ancient Shiloh. It was the center of Israel's worship before Jerusalem was conquered by David's men. The tent of meeting with the ark of the covenant rested there for over three hundred years. David, after capturing Jerusalem, arranged to move the ark to Jerusalem.

Speaking of Jerusalem, you will soon begin to see her sparkle on the distant horizon.

We are now flying over the Mount of Olives, and to the southwest is the Valley of Hinnom. You can see the ongoing fires on the north slope of the valley. That is where Gog and his prime minister/false prophet were thrown seven weeks ago. The fires will continue as a witness for all to see and consider.

We are now over Mount Herzl and over there is the Shrine of the Book. This museum and its precious holdings, the Dead Sea Scrolls and the *Aleppo Codex*, survived seven years of Gia Union rule and Gog's attempt to wipe out Judaism, including his final attack on the city seven weeks ago.

"We are now coming to a landing on the grassy slope adjacent to the Joppa Gate. Here we are! We'll walk from here."

Passengers disembark and assemble around Yoni.

"Hold up! wait a minute everyone. Let me introduce you all to an old friend. This is David."

David walks over and gives Yoni a big hug. Then he greets everyone and addresses the group.

"This is my long-standing friend. We used to hunt and travel together. We even fought the Philistines together. And we had to hide from his dad when he became angry with me. Yoni was called Yonatan back then. You would know him as Jonathan. He's modified and modernized his name a bit now."

Yoni says, "Hey guys, today is 'Bible come alive day' for you. And I mean it literally. This young, tough looking man is my friend David, King of all Israel. Again. May his reign be long! And it will!"

Rivka can hardly believe her eyes. "You mean this is *the* real David and *you* were the good friend of David? You are the Jonathan we read about in the Bible?"

Yoni is modest. "Well, it would come out sooner or later. But that's enough about me since we'll have a lot of time together. But today, right now, you're meeting King David, for real. And you're going to meet a lot of other familiar figures, not make-believe, not actors. This is all for real."

Raquel looks at David. "But we would think that you should be sitting up on a golden throne or something, not just walking around the Joppa Gate."

"Well, the throne gets hard after a couple of minutes and besides, don't you know that most of the business is conducted at the gates? Check it out in your Bible.

"Hey guys, it's good to meet you, but you have a busy schedule and should be getting along. I believe you have an appointment with Yeshua.

"Shalom!" David heads back toward the Gate.

Teah is almost breathless. "Thanks, Yoni! What can I say now?"

"*B'vakasha*, you're welcome. Let's get moving on. Wait a minute . . . Hey, Sarah! Come over and meet my friends," Yoni calls.

Over comes an attractive young redhead with baby blue eyes.

"Shalom, welcome to Jerusalem," she says softly in perfect English. "I hope Yoni is giving you a good tour. Where are you all from?"

Miriam says, "Six of us are from Wisconsin, others from other places in America, some from the Caribbean, and I am from Australia."

"I hope you all get to meet my husband Abraham. He is obsessed with hospitality. But he is tied up right now trying to set up an old fashioned welcoming center down in the Judean Desert. He wants to greet people like he did three thousand years ago.

"If you get to go down toward Jericho stop by to meet him. He will give you a big meal, a lot of old stories, and a camel ride through the desert.

"Personally, I think he just likes the desert and riding on a camel."

Rivka says, "Wow, really! You're Sarah, the Sarah, Abraham's wife, mother of Isaac, and grandmother of Jacob and Esau?"

"That's me," Sarah replies smiling. "That life was surely an adventure but it's really much better today. Now starts the biggest adventure, seeing what all those grandkids are up to presently."

Ellis leans over to the other guys. He whispers, "I think I understand the story of Abimelech better now."

"Who's Abimelech"? Ian asks. Ellis replies, "Check your Bible." The men chuckle, the ladies shake their heads in displeasure.

"We're on our way to meet Yeshua in the Temple. Have you seen him around lately?" Yoni asks.

"Yes," Sarah says. "I bumped into him in the marketplace a few minutes ago. I guess he knows you're coming?"

"Yes, so we better get going."

"Nice to meet you Sarah, shalom," they all say in unison. "Shalom."

Yoni now has the whole group of six hundred assembled in four side-by-side lines winding through the narrow streets. He resumes his ongoing commentary as he shepherds the group through the old Armenian quarter.

"You know the city is still an old city with Turkish architecture and narrow streets. A rebuilding plan is underway but you'll hear more about that later. Building the temple and its precincts is the highest priority. And even that is still under construction. There's a lot of activity here, so watch your heads."

Just above them they hear a voice. "Shalom, Yoni!"

They all look up and there, seemingly suspended, is a man with wings, who seems at first impression to be an angel . . . and the last impression remains unchanged. He comes to a standing landing just ahead of Yoni.

"Shalom, Gabriel!" Yoni says. He gives him a big hug while avoiding wing contact. "Gabriel, I want you to meet our new friends who have moved to Israel. This is their first trip to Jerusalem, I think. Have any of you been to Jerusalem before?"

About a hundred raise their hands.

"Well, first time since moving to Israel, that I know for sure."

Gabriel says, "Either way it's good to meet you all, and welcome home."

"Mr. Gabriel?" The voice of a young girl comes from the back. "Do you play your trumpet every day and practice like I used to have to do, or do you play only on special occasions?"

"Well, Anna . . . on special occasions. But stick around, tomorrow will be Rosh Hodesh, the new moon, and you'll hear me play loud and clear. But you should keep practicing. You see, I may not practice every day now, but I have been blowing the trumpet for a long time."

Anna is overwhelmed. "Thanks," she quietly says.

"We're on the way to the temple to meet Yeshua. Have you seen him?" Yoni says.

"Yes, I just saw him in the marketplace a few minutes ago, does he know you are coming?"

"Yes."

"Shalom."

"*Shalom, l'heit'ra'ot,* See you later!"

The group then makes a turn to go through the old marketplace, a short cut to the Temple. As they are crowding through the narrow streets, there is a call, "Shalom, Yoni!"

"Yeshua!" Yoni calls back.

Beaming, Yoni turns back to the group. "Well my friends, here is Yeshua!"

Everyone turns to see, trying to absorb the magnitude and significance of this incredible experience. All had eagerly anticipated the event and had tried to visualize what it would be like. No one was disappointed. The Yeshua they beheld could only be described as a flawless embodiment of goodness, love, and perfection. It was strange, but clear, the beautiful stereotyped images developed over the ages, by historians, great artists and followers, had all failed to capture his magnificence.

When Yeshua first spoke, the group was surprised by his direct, ordinary, and distinctly "human" manner. First he looked to Yoni and asked, "Well, these must be the folks I am to meet?"

"Yes," Yoni answers.

Yeshua turns to face the group. "Sorry I'm late, got tied up in the marketplace looking for something special for Miriam (Mary), my mother."

Yeshua goes over and begins to greet each person by name, giving each of them a big bear hug and engaging in some words of comfort unique to each one. This takes, of course, some time. But no one is in a hurry. He seems to know everyone, not only by name and place of origin, but by their special gifts and concerns. It doesn't take long to bond with him. No words are wasted. Every word is significant.

When Ellis' turn comes, he looks Yeshua in the eye and says, "There's one question I have been waiting to ask you."

"Okay."

"Have you ever been here before?"

Yeshua breaks out into loud laughter. After a minute or two he responds, "Ellis, that joke is as old as the hills, but you are the first to actually use it on me. The answer is, yes!"

"Thanks!" Ellis says.

Joel leans over to Isaac. "What was all that about, Isaac?"

Isaac whispers, "It is an old Jewish joke."

"Thanks." Joel lightly punches Ellis on the shoulder.

Yeshua continues to greet everyone personally and bonds with each one.

"Jacob!" Yeshua says. "Thanks for bringing our friends from the Amazon."

Yeshua turns to the Amazon chief and speaks in the Tapi language. "Ubiratan, your name means 'freedom.' It is a fitting name for your people's special place in this new world. And accept my thanks for taking care of Jacob and his cousins. Welcome to Israel and Jerusalem.

"Jacob, thanks also for bringing home more missing pages of the *Aleppo Codex.*

"Tom and Bart, I appreciate what you and the other Seventh-Day Baptists did for me in Beloit."

Tom asks, "Will we be punished for accepting The Marck?"

Yeshua gives him another a big hug. "You already have been. You are forgiven. It is over. Let's focus on the future."

Bart then asks, "I'm finding all this a little confusing. What shall I call you? Jesus or Yeshua?"

Yeshua kindly smiles. "Over the centuries, I have been called many names by many peoples and I accept and respond to all. The name on your lips is not as important as the intent. But, I prefer Yeshua."

Bart is still perplexed. "What about the other members of my family and my friends? Where are they? If they are alive, will I ever see them again? Is my homeland also being rebuilt? There are too many unanswered questions!"

Yeshua replies, "I understand your concerns but I assure you, I am in control. You should recognize that the future will not resemble the past, and your 'now' does not predict the future. The answers will come as the perfect order of our new world unfolds."

Yeshua now pauses, looks to the crowd and points to the west. "Here, come walk with me as we are heading in the same direction. We have a lunch set up for you in the courtyard. I will give you a tour of the Temple afterward. But you should not go into the Holy of Holies. That's for me, the king and high priest, only.

"Although you don't know it, about a third of you are Levites and a many of you are cohans (priests), so you will soon serve here too. We have a lot to talk about. This is great. I have waited a long time for this!

"Welcome home! Welcome home, everyone!

"And, Teah, I'm glad to hear that you will soon be moving to Jerusalem to assist us with touring organization and help make sure everyone gets the most out of their excursion."

"Me?" Teah replies with surprise.

Yoni flushes with embarrassment. "I haven't told her yet."

After Teah has time to catch her thoughts, she stands up straight and looks at Yeshua. "Of course. I'm always at your service."

Isaac looks puzzled and maybe a bit concerned. He avoids making eye contact with Teah or with Yoni.

They are now entering the Temple complex. There are forty mikvahs, immersion pools, just outside the temple's southern gate. The water is cool and refreshing. It comes directly from the river flowing out from under the altar of the temple.

With the exception of Bart and Tom and a few others who are unfamiliar with Jewish practices and customs, everyone immediately immerses. Being Baptists, the two are unfamiliar with and a bit skeptical of the mikvahs. Bart complains to Tom, "I don't understand this. We have both been baptized in our church, why are we expected to get into these?"

Yeshua suddenly appears behind them. "Immersion in a mikvah has been a Hebrew practice for thousands of years. The practice of baptism developed from these Hebrew customs. I respect your original baptism, but there are other reasons for immersion. In this case, it relates to honoring the temple. Again, I remind you . . . answers will come as the perfect order of our new world becomes known."

Tom and Bert, along with some others, hesitantly immerse themselves into the mikvahs and, still dripping-wet, join the others entering the temple courtyards.

Yeshua now addresses the group. "Sorry for the construction inconveniences, it will be going on for some time. But at least you do not have to wear those silly yellow hard hats!

"Once the core of this complex is finished, I am going to raise the base of this mountain by one thousand cubits. That is 1,500 feet to you.

"You see the river of water that now flows from under the altar? It's a gusher, don't you think? It flows into the Kidron Valley and then through the new Olivet Valley. It joins the Wadi Kidron. It is on its way to the Jordan and into the Dead Sea. It is a start for the healing of the Judean wilderness and the Dead Sea itself. But the guys who invested in Dead Sea mud and cosmetics are still a bit skeptical." His delayed smile verifies his humor.

"That cable car across the valley to the Mount of Olives will be going the other direction." He smiles again, and pauses. Some take a while to catch on. There are the delayed smiles and chuckles.

"After I raise this Temple Mount, more water will flow out and around the city's southern border, then down by Beth Shemesh through the Sharon and into the Mediterranean.

"Tributaries will flow out to the north and the south. A big branch will turn south into the Negev. All this water is living water, coming from under bedrock, from under the altar. It will be a mammoth gusher, looking a lot like the Niagara Falls in North America.

"It will give life to the land.

"We're going to irrigate this land like it has never been in five thousand years. It will become truly a land of milk and honey. It will be filled with villages, farms, orchards, parks and waterways. And there will be people, happy people with children, playing in the plazas.

"Now, I think you understand that I could do this all by myself. Raise the mountain, open up the fountains, then rebuild the roads, plazas, the cities and villages, the farms and orchards, and re-forest the land. That would be the easy way. But I want to share the fun and joy of *'creation'* and transforming this land with you. I am so thankful that you have joined me in this great project. You won't be sorry.

"Some of you are trained in engineering, biology manufacturing, communication, hydrology, transportation chemistry, engineering, genetics, physics and construction Others of you will need some new or additional training. Everyone is going to need refreshers and updates. Education will be a major part of our program. What you need, you will get, so that we can get this job done and done right for a change.

"Now after we heal and rebuild this land, we will march forward while spreading this way of life throughout the world. As I have explained to some of you, the future will not be like the past, and today's 'now' does not limit what is ahead. I know you have many questions. And all will be answered. As we join together to heal the whole world, everyone has a special role in this adventure.

"Oh, I'm sorry . . . forgot to serve lunch.

"You see, I get carried away with all of this. Let's have lunch, then we can get on with...

Rebuilding the Planet!"

To be continued

For full-color high-definition viewing of illustrations in this volume, other illustrations, future episodes, further readings and links, Rebuilding the Planet blogs, and more, go to: www.RebuildingthePlanet.com